To my won_____ _____,
Karen; the _ ♡ S0-BXZ-560
    Inquisitive scientist u
        Know!
Here's to many more
        scuentitic adventures!
            Love,
                Sue

A BOOK OF
SCIENTIFIC CURIOSITIES

Also by Cyril Aydon

Charles Darwin

A BOOK *of*
SCIENTIFIC
CURIOSITIES

Everything You Need To
Know About Science – But
Never Had Time To Ask

———————

CYRIL AYDON

CARROLL & GRAF PUBLISHERS
New York

Carroll & Graf Publishers
An imprint of Avalon Publishing Group, Inc.
245 W. 17 th Street
New York
NY 10011-5300
www.carrollandgraf.com

AVALON
publishing group incorporated

First published in the UK by Constable,
an imprint of Constable & Robinson Ltd 2005

First Carroll and Graf edition 2005

ISBN-13: 978-078671-628-9
ISBN-10: 0-7867-1628-2

Printed and bound in the EU

CONTENTS

THE HEAVENS

THE EARTH

THE LIVING WORLD

MASS AND ENERGY

THE NATURE OF MATTER

SCIENTIFIC INNOVATORS

SCIENCE IN SOCIETY

KEEPING COUNT

APPENDIX 1: MEASURING THINGS

APPENDIX 2: TIMELINES

ACKNOWLEDGEMENTS

Many people have helped with the writing of this book. My biggest debt is to Jim Honeybone, who has read not only every word printed here, but many others that did not survive his critical scrutiny. Deborah and Sue Aydon, Mike Fenner and Jean Honeybone read portions of the text, and forced me to think hard about what I was trying to achieve. Jean Button, librarian of the Warriner School, Bloxham, provided several helpful pointers. I am especially grateful to my wife Joyce for her personal support and her technical expertise, both of which played an essential part in the writing of the book. Among people whom I have not personally met, I owe a particular debt to Isaac Asimov, whose *Biographical Encyclopaedia of Science and Technology* made the task of writing the book much easier than it would otherwise have been.

This is my second foray into popular science. The first, my biography of Charles Darwin, owed its publication to the faith and support of my editor at Constable & Robinson, Carol O'Brien. No one could have had a wiser and more supportive editor, and I count myself lucky that I had the benefit of her advice during the writing of this book also. Her final act of kindness before retiring was to commit me to the skilled care of Helen Armitage, who has overseen the book's production. The task of ensuring the quality of the finished text fell to Claudia Dyer and Penelope Isaac; and I am indebted to them for their many helpful suggestions.

MY SCIENTIFIC CURIOSITY

When I was young, and it was too wet to go out to play, I would wander into the front parlour and lose myself in a book. Most of the books there were adventure or detective stories, which owed their presence in the house to their non-return to local shopkeepers' circulating libraries. But among them were several volumes of *Harmsworth's Self-Educator*, a self-improvement manual issued in weekly parts, and accumulated over a period of about seven years by an aspiring uncle. Among the instalments explaining the mysteries of fretwork, dressmaking, and double-entry bookkeeping, were others devoted to astronomy, geology, and natural history. They were old when I discovered them; and the knowledge they contained was no doubt already out of date. But for a young boy meeting such subjects for the first time, it was like opening Aladdin's cave; and it was the beginning of a lifetime's fascination with the history of science. Looking back, I realize that none of my despairing teachers had any idea of my secret store of extracurricular knowledge, and it never occurred to me to mention it.

My later studies lay outside science, but the fascination with science history never left me; and the pleasure it has given me has been immense. This book is an attempt to give something back, by retelling some of science's most intriguing stories; and by passing on some of the astounding facts that scientists have uncovered about our own history, and about the universe we inhabit.

The book can be read straight through, as an introduction to 2,000 years of scientific discovery. But readers who prefer to treat it as a 'lucky dip' will, I hope, find interest wherever they open it.

The book does not assume any prior knowledge of science, or of mathematics. And there is only one formula in the book: Einstein's $E = mc^2$. It would not be a history of science if that one were left out.

COUNTING IN SIXTIES The only reason for counting in 10s is that we have 10 fingers. Apart from that, there is nothing special about 10. Digital computers count in 2s, because that suits the way they work. Four thousand years ago, the Babylonians counted in 60s. The number we write as 150, they wrote as II<<< (two 60s and three 10s).

As a basis of counting, 60 has a lot going for it. It is divisible by 2, 3, 4, and 5. (It is the only number below 120 that is.) This makes it handy for measuring things we want to divide up into smaller parts. The Babylonians' hour was not the same as ours; but it was they who gave us 60 minutes to the hour, and 60 seconds to the minute. And it was they who divided a circle into 360 degrees (6×60), and a degree into 60 minutes.

The disadvantage of counting in 60s, rather than 10s, is the burden it places on the memory. It is fairly easy to learn up to 10 times 10 by heart; but memorizing tables up to 60 times 60 would be a tall order. The Babylonians got round that problem by using written tables. But it did rather slow down the arithmetic.

THALES OF MILETUS Western science was born in Ancient Greece, and in the Greek-speaking cities around the Mediterranean. The Greeks considered *their* science to have begun with the philosopher Thales. He was born in Miletus, a city on the Aegean coast of Asia Minor, in 624 BC; and he died there in 546 BC. It is hard to say how many of the achievements with which he was credited were his own idea, and how much he owed to knowledge acquired on his travels in Egypt and Mesopotamia. But when his contemporaries drew up their list of 'The Seven Wise Men', they placed Thales first.

His interests were wide-ranging. He was the first person we know of who made a serious study of magnetism. Most importantly, he laid the foundations of the Greek system of deductive mathematics that culminated two centuries later in the work of Euclid. The philosopher Aristotle, who lived much later, tells two stories about him that may or may not be true; but if they aren't, they deserve to be. In response to the old jibe 'If you are so clever, why aren't you rich?' he entered into contracts to hire a large number of olive presses at low rentals, at a time when his study of the weather told him there would be a bumper crop. There was, and he made a killing, hiring his presses out. It was also claimed (and perhaps this should be mentioned on every Business Management course) that he fell into a pit while stargazing.

PYTHAGORAS AND HIS ODD IDEAS   Pythagoras, who was a younger contemporary of Thales, was born around 560 BC, on the island of Samos, just a short sail from Thales' home in Miletus. He is remembered now mainly for his theorem about the three sides of a right-angled triangle; but for hundreds of years after his death his teachings were a dominant influence on the thinking of mathematicians, scientists, and moral philosophers. Many men of genius have harboured strange superstitions while maturing brilliant insights, but for sheer loopiness, some of Pythagoras' ideas would be hard to beat. His followers were forbidden to eat beans on the grounds that a bean, if buried in the ground for 40 days, and covered with dung, would assume human form. He was a believer in the transmigration of souls, so that a man's soul might, in a previous existence, have inhabited the body of a jellyfish.

But if Pythagoras' speculations led his disciples into a thicket of superstition, the mathematical and astronomical insights with which he is credited left later scientists heavily in his debt. It was Pythagoras who made mathematics into a unified logical system, rather than a set of rules for special cases. He was also the first person who is known to have speculated that the Earth might be spherical in shape. Neither the Babylonians, nor the Egyptians, nor the earlier Greeks, had been aware of the Earth's true shape. Homer had thought of it as a convex disc, surrounded by a river. Some of his contemporaries believed it to be a plate, supported by four elephants standing on a turtle. Whether or not Pythagoras was actually the first to hit upon the truth, it was he who introduced to astronomy the image of a globe suspended in space that was the foundation of future progress in the science.

One of the most remarkable achievements of his school was the discovery of the mathematical basis of musical pitch. Lots of people must have noticed that a short string gave out a higher note than a long one. It was Pythagoras who discovered the mathematical relationship between the length of a string and the note it emitted, so that if the length of a string was doubled, the sound dropped by an octave; if the ratio of the lengths was three to two, the difference in pitch was a fifth, and so on.

ARISTOTLE One of the troubles with really great thinkers is that their opinions can possess men's minds long after they are dead, to the extent that they get in the way of new thinking. This was certainly the case with Aristotle. Nearly 2,000 years after his death, it was still standard practice among scholars to settle arguments by asking, 'What did Aristotle say?'

He was born in Northern Greece in 384 BC, the son of the court physician to Amyntas, King of Macedon. Between the ages of 17 and 37, he lived in Athens, where he was a member of the Academy, and Plato's star pupil. After Plato's death, he travelled for a dozen years. In 342 BC, when he was 42, he was summoned home by Philip II, Amyntas' successor, to tutor his 14-year-old son, the future Alexander the Great. Six years later, Philip was assassinated, and Alexander succeeded to the throne. He went off to conquer the world; and Aristotle returned to Athens, where he founded a school of his own: the Lyceum.

Aristotle concerned himself with the whole of human experience, including science and logic, ethics, politics, and even literary criticism. It was in natural history that he made his most important discoveries. He was one of the greatest biologists of all time. His classification of the invertebrates (animals without backbones) was superior to that of Linnaeus 2,000 years later. He minutely studied 500 animal species. Amazingly for a Greek (Greek gentlemen did not usually engage in manual labour) he dissected 50 of them. He drew up a hierarchy of living forms, embodying the idea of a 'Chain of Being', from lower to higher forms. This did not, however, lead him to support the theories of evolution espoused by some of his contemporaries. For him, the essence of both the animate and the inanimate worlds was their unchanging perfection.

One area in which Aristotle had an unfortunate influence on later generations was cosmology. It was he who introduced the idea of a set of concentric heavenly spheres that revolved around an immobile central Earth, and to which the Sun and Moon – and all the planets – were affixed. It was an image that would prove difficult to escape from, and it got in the way of clear thinking about the nature of the cosmos for hundreds of years after his death. Even when Nicolaus Copernicus published his model of a universe in which the Sun, rather than the Earth, was the centre, his planets and stars were still attached to Aristotelian spheres. But it should be said in Aristotle's defence that this was a problem that owed more to his followers, and their distorted conception of his teaching, than it did to the man himself.

Great naturalist though he was, Aristotle could be a fallible guide, as later generations would discover. One of his many confident pronouncements was that women had fewer teeth than men. We will never know whether his mistake was the result of his having counted Mrs Aristotle's teeth, or his having failed to do so.

PYTHEAS THE EXPLORER   Pytheas lived in Massilia – the modern Marseilles – around 300 BC. In his day, the Mediterranean shores were lined with Greek colonies, of which Massilia was the most westerly. Pytheas was an intrepid traveller, and later geographers drew heavily on his work. But his descriptions of distant countries seemed so far-fetched to his contemporaries, who knew only the Mediterranean, that they dismissed him as a fantasist. Unfortunately, his most famous work, *On the Ocean*, has not survived, and we are dependent on second-hand accounts for both his descriptions and his ideas.

On his most daring journey, Pytheas explored the coast of northwest Europe. He visited Britain, from where he brought back descriptions of drinks made from grain and honey, and a country he called Thule ('Toolay'), which may have been Norway. Prevented from travelling farther north by fog, he proceeded to explore the Baltic as far as the mouth of the Vistula. One of his descriptions that gave rise to particularly ribald disbelief was of a northern sea made stagnant by 'a mixture of air, earth, and water'; but to anyone familiar with pack ice, it does not sound far-fetched.

Pytheas was not just an explorer, he was a true scientist. His most notable achievement was his explanation of the tides, a phenomenon virtually unknown to his contemporaries living around the almost tideless Mediterranean. Pytheas attributed the tides to an attraction emanating from the Moon, a proposition that reinforced his image as a fantasist. The world had to wait until Isaac Newton's *Principia* (1687) for the proof that, on this point at least, Pytheas had been absolutely right.

THE MOON AND THE TIDES   The Moon is not the only influence on the tides. The Sun also plays a part. But the Moon's tide-raising power is twice as great as the Sun's. Given how much bigger the Sun is, this may seem strange. But the tides are caused by the *differential* effect of gravitation on the body of the Earth and on its surface water. Because the moon is so close to the Earth, the *difference* between its pull on the solid Earth and on the surrounding seas is greater than the corresponding difference in the case of the Sun.

There is always a high tide on the side of the Earth nearest the Moon, and another on the opposite side. This is why most places have two high tides a day. It seems as though the tides are sweeping around the globe; but what is really happening is that the Earth is revolving under the tides.

The Moon makes one complete journey around the Earth every four weeks. In many places, higher tides than usual ('spring tides') occur once a fortnight, around the times of new moon and full moon, because the Sun, the Moon, and the Earth are then more or less in a straight line, and the two tide-raising forces are pulling in the same direction. Much lower high tides ('neap tides') occur halfway between these dates – at the Moon's first and last quarters – when the Moon's gravitational force is operating at right angles to that of the Sun.

Exceptionally high tides are associated with a full or new moon around the time of the equinoxes, in late March and late September. This is because, at the equinoxes, the Moon's orbit around the Earth intersects with the plane of the Earth's orbit around the Sun, causing the three bodies to be in a straight line, not only when viewed from above, but from the side as well.

The frequency and height of the tides depend on the configuration of the adjoining land. The English port of Southampton has four tides a day. Some places in the China Sea have only one. The greatest tidal range – 15 metres/50 feet – is met in the Bay of Fundy, off Nova Scotia. By contrast, the difference between high and low tides in the Mediterranean, and around some Pacific islands, is less than 0.5 metre/2 feet.

OUR NEIGHBOUR THE MOON The Moon is our nearest neighbour. Compared with the Sun, or any of the planets, it is *very* near. The Sun is 150 million kilometres/93 million miles away, but the Moon's average distance is only 385,000 kilometres/239,000 miles. Because the Moon's orbit around the Earth is elliptical, not circular, the actual distance varies between 350,000 and 400,000 kilometres/221,000 and 253,000 miles. Its diameter is only a quarter of the Earth's: 3,500 kilometres/2,200 miles, compared with 12,700 kilometres/8,000 miles. It is a giant mirror, which owes its brightness to sunlight reflected from its surface rocks. It is not a very efficient mirror. Its *albedo* – its reflecting power – is only 7 per cent. So only 7 per cent of the sunlight falling on its surface gets reflected back into space. But the Sun's light is so bright that – when the Moon is full – a mere 7 per cent of it, bounced back for a quarter of a million miles, is still enough to find one's way home by.

The Moon always presents more or less the same face to the Earth, because the time it takes to revolve once on its axis – $27\frac{1}{2}$ days – is the same as the time it takes to go round the Earth. This is not a coincidence. It is the result of millions of years of gravitational drag between the two bodies. It is not precisely the case that the Moon always presents the same face to the Earth, because it is subject to an oscillating motion called *libration*. As a result, about 60 per cent of the Moon's surface is visible from the Earth at one time or another, and about 40 per cent is permanently invisible.

The interval between successive New Moons is $29\frac{1}{2}$ days, not $27\frac{1}{2}$, because while the Moon is orbiting the Earth, the Earth is going round the Sun. It takes the Moon an extra two days to reach a point where the three bodies are once again in a straight line.

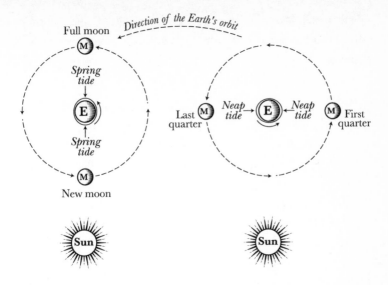

Figure 1. The Moon and the Tides
New moon and full moon occur when the Earth, Sun, and Moon are in a straight line. Tides are not so high at the first and last quarter, because then the Sun and Moon are pulling at right angles to one another.

THE EARTH'S SPIN The friction between the tides and the ocean bed slows the rate at which the Earth spins on its axis. As a result, the day is getting longer – by about $\frac{1}{1,000}$th of a second per century. There is a physical principle called 'The Conservation of Angular Momentum' that requires the Moon to move farther away as the Earth slows down – which it is doing, by about 4 centimetres/1.5 inches a year.

This may not sound much, but it has been going on for a very long time; which means that the Moon and the Earth must once have been much closer together – and spinning much faster. But whether they once formed a single mass, and whether the Moon was created as the result of some cosmic collision, are questions to which science cannot yet give a confident answer.

ACCURATE CLOCKS Tiny fractions mount up. There are more than 30,000 days in a century. So although each day is only $\frac{1}{1,000}$th of a second longer than the corresponding day 100 years earlier, the accumulated slowing-down amounts to about 30 seconds. This means that, after 100 years, a clock synchronized with the daily revolutions of the Earth would be 30 seconds slow compared with one that measured time consistently. To avoid this, the world's clocks are put forward by 1 second once every few years. We are not aware of this, because the change is made to the reference clocks from which all other clocks take their time.

In addition to this long-term slowing-down, the Earth's speed of rotation varies slightly, due to a number of short-term influences. It may still be an adequate timekeeper for everyday purposes, but modern technology needs greater precision than the Earth can provide; and the world's clocks nowadays take their time from the movements at the heart of the caesium atom, not from the movement of the Earth.

ARCHIMEDES OF SYRACUSE Archimedes was the greatest scientist of ancient times. Some people consider him one of the greatest of all time. He was born in Syracuse, in Sicily, in 287 BC. Syracuse already had a history of 500 years of civic life; and in his day it was a powerful, independent state, whose rule extended over much of southern Italy.

Archimedes is remembered as the person who supposedly leapt from his bath, and ran naked into the street crying *'Eureka!'* ('I've found it!'). What he had found was not the soap, but the *principle of flotation*: the principle that explains why iron ships don't sink. He may well have had the idea in his bath, but there is no evidence to support the rest of the story. He was an aristocrat – the son of an astronomer, and the confidant of his king – and we may safely credit him with more concern for his dignity than the traditional story would allow.

One thing that marks Archimedes out from other scientists of his time is that he was both a mathematical genius and a gifted engineer. It was he who laid the foundations of the science of mechanics, including the principles governing the action of pulleys and levers. In many of his discoveries it is hard to say whether the source of his success was his mathematician's insight or his engineer's intuition; and it is fitting that the saying for which he is best known should be, 'Give me a place to stand on, and I will move the Earth.' Despite the brilliance of his many mechanical inventions, he placed so little value on them that he left virtually no record of them in his own writings. In his own view, it was his mathematical work that counted, and it was to the furtherance of the study of mathematics that his writings were dedicated.

Archimedes' most important work was in geometry. Among the many problems he successfully tackled were those concerned with finding the areas of curved shapes. He also arrived at an accurate figure for the value of *pi*. And in his time off from revolutionizing the study of geometry, he established the foundations of the future study of mechanics, of statics, and of hydrostatics. It has even been suggested that, if he had had access to a better system of numerical symbols, he might have beaten Newton to the invention of calculus.

Archimedes met his death at the hands of a Roman soldier. The ruler of Syracuse had broken an alliance with Rome and thrown in his lot with the Carthaginians, whose general, Hannibal, was carrying all before him. The Romans sent a fleet to attack the city, and after a three-year siege, which Archimedes' war machines helped to prolong, they conquered it. The Roman general Marcellus held Archimedes in high regard, and gave instructions that he was to be brought to him unharmed. But the soldier who discovered him engrossed in a mathematical problem lost patience at his slowness to respond, and ran him through.

In 1453 – nearly 1,700 years after his death – when another Mediterranean city, Constantinople, fell to an invading force, a handful of Greek scholars sailed west, carrying the only surviving copies of some of the treasures of Greek science. These included some of Archimedes' writings. By a minor miracle, they fell into the hands of a German astronomer, Regiomontanus, who put in train a programme of translation that continued after his own death. It was partly by this lucky chance that Archimedes' insights survived, to provide the foundation for the scientific revolution initiated in the sixteenth century by Copernicus and Galileo.

WHY IRON SHIPS FLOAT Archimedes' principle is one of the easier scientific laws to understand. It states simply that floating bodies displace their own weight of the liquid in which they float. An iron bar sinks, because it weighs much more than the volume of water it displaces. An iron ship floats, because

the Sun – with the stars as a distant backdrop – and it became difficult to believe in a 'science' that talked of the Sun being 'in Leo', or 'entering Sagittarius'. It became even more difficult to believe in one that claimed to predict the future from the relative positions of planets millions of miles away. Some astronomers did continue to study astrology but, after 1700, the science of astronomy and the pseudo-science of astrology took different paths. Now, of course, astrology is just a branch of the entertainment industry. But so long as it continues to sell newspapers, its future is safe.

PATHS ALONG THE ECLIPTIC   During the course of a day, the Sun seems to follow a path across the sky. It doesn't, of course. The Earth is revolving on its axis, from west to east, and this makes the Sun appear to be travelling from east to west. The Sun has another motion that is just as illusory. Anyone who makes a habit of studying the sky just after sunset, as the stars begin to appear, will notice, as the weeks go by, that the Sun seems to set in a different part of the sky. Over the course of a year, it seems to perform a complete circle of the heavens, against the background of the stars. The path of this journey, plotted on a map of the stars, is called the ecliptic. But, like the constellations, the Sun's journey along the ecliptic is an optical illusion. What is really happening is that the Earth is travelling around the Sun, like a child on a carousel. Just as, to the child's eyes, a person standing in the middle of the carousel seems to move against the background of the funfair, so, to an observer on the Earth, the Sun seems to move against the background of the stars.

Because the Moon's orbit around the Earth, and the orbits of the planets around the Sun, all lie in the same plane, they also appear to follow a path along the ecliptic.

THE ZODIAC The band of stars through which the Sun, the Moon, and the planets appear to travel is called the zodiac (Greek for 'circle of animals'). It is divided into 12 constellations, which we inherited from the Greeks. They have names such as Leo (the Lion), Taurus (the Bull), and Cancer (the Crab) that supposedly reflect their appearance. The constellations of the zodiac, with the dates when the Sun 'enters' and 'leaves' each one, are as follows:

Aries (the Ram)	21 March–20 April
Taurus (the Bull)	21 April–21 May
Gemini (the Twins)	22 May–21 June
Cancer (the Crab)	22 June–23 July
Leo (the Lion)	24 July–23 August
Virgo (the Virgin)	24 August–23 September
Libra (the Scales)	24 September–23 October
Scorpio (the Scorpion)	24 October–22 November
Sagittarius (the Archer)	23 November–22 December
Capricorn (the Goat)	23 December–20 January
Aquarius (the Water Carrier)	21 January–19 February
Pisces (the Fishes)	20 February–20 March

Because of the phenomenon of *precession*, the signs of the zodiac no longer correspond to the Sun's position in the sky on these dates. For most of the period from 21 March to 20 April the Sun is nowadays in Pisces, not Aries.

STELLAR MAGNITUDES   There are about 6,000 stars visible to the unaided eye. About 2,000 can be seen at one time from any point on the Earth's surface. The brightest are the first magnitude stars – the name given to them by Hipparchus, the greatest astronomer of ancient times. Hipparchus, who was born in northwest Turkey around 190 BC, built an observatory on the island of Rhodes. It was he who, in 129 BC, completed the earliest known catalogue of the stars, and who divided the naked-eye stars into the six categories, from first magnitude to sixth, that we still use today.

Hundreds of years later, in 1856, his scale was standardized by the English astronomer Norman Pogson. Pogson defined Magnitude 1 as being 100 times brighter than Magnitude 6, making each magnitude 2.51 times brighter than the next faintest. (2.51 $\times$ 2.51 $\times$ 2.51 $\times$ 2.51 = 100). A star of Magnitude 1 is $2\frac{1}{2}$ times brighter than a star of Magnitude 2, and a star of Magnitude 3.5 is $2\frac{1}{2}$ times brighter than one of Magnitude 4.5. A very bright star might be Magnitude 0 ($2\frac{1}{2}$ times brighter than Magnitude 1), or even *minus* 1 ($2\frac{1}{2}$ times brighter than Magnitude 0). These *apparent* magnitudes, as astronomers call them, are no indication of the size of a star, or of its real brightness. Some bright stars really are exceptionally big; but some only appear so because they are closer to the Earth than others. An example of a star that is both apparently and truly big is Betelgeuse ('beetle-juice'), in the constellation of Orion. Betelgeuse (Magnitude 0.5) belongs to a class of star called 'red giants'. It is so big that, if the Sun were placed at its centre, the Earth's orbit could be contained within its circumference.

SOME BRIGHT STARS

Star	Magnitude	Distance from Earth (light years)
Sirius	-1.5	9
Canopus	-0.7	74
Arcturus	0	34
Capella	0.1	41
Rigel	0.1	815
Antares	1.0	220
Polaris (the Pole Star)	2.0	430

The brightness of any light-source falls off according to the *square* of its distance away (Figure 7). If two identical candles are placed so that one is *twice* as far off as the other, the nearer one will appear *4 times* brighter. The same is true of stars. Rigel and Capella appear equally bright. But Rigel is 20 times farther away than Capella. If Rigel were as close as Capella, the Earth would receive 20 × 20 = 400 times as much light from it, and it would look 400 times brighter than it does. (Its real luminosity is 60,000 times greater than that of the Sun.)

LIGHT YEARS Stars are so far away that our usual measures of distance can't handle the numbers involved. So astronomers have come up with a handier measure: the light year. A light year is a measure of distance, not time. It is the distance travelled by a ray of light in one year. Light travels 300,000 kilometres/186,300 miles in a second, which is 9.5 million million kilometres/6 million million miles in a year. So when we say that Rigel is 815 light years away, we are saying it is 815 *times* 9.5 million million kilometres/6 million million miles away, which is *quite* far.

THE SEASONS The Earth is a like a giant gyroscope. The tilt of its axis remains the same during its annual journey around the Sun. For part of the year the northern hemisphere is tilted towards the Sun, and the southern hemisphere is tilted away.

At other times the position is reversed. As a result, the number of daylight hours in any given place, and the heat of the Sun's rays, vary during the course of a year. The nearer to the poles, the greater is the effect. This is why regions away from the equator have seasons with such marked differences in temperature. In Miami, in the middle of February, the Sun's rays have 80 per cent of the heat they have in June. In Fairbanks, Alaska, the ratio is only 20 per cent. Near the equator, temperature fluctuations are less marked, and the seasons are differentiated from one another by varying amounts of rainfall rather than varying amounts of sun.

NATURAL TIME The story of clocks and calendars is the story of humanity's attempt to reconcile two different concepts: *natural time* and *artificial time*. Natural time is based on the movements of the Sun, the Moon, and the stars. Artificial time is arbitrary, and has nothing to do with astronomical phenomena.

Every calendar system ever devised is built around days, months, and years. These reflect three natural phenomena: the revolution of the Earth on its axis; the revolution of the Moon around the Earth; and the revolution of the Earth around the Sun. It wasn't necessary for early calendar-makers to know that

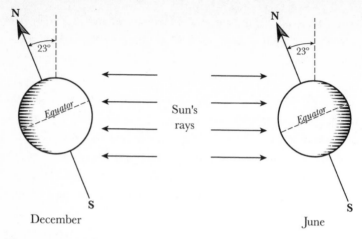

Figure 2. The Earth's Tilt and the Seasons
In December, the northern hemisphere is tilted away from the Sun.
In June, on the other side of the Earth's orbit, it is tilted towards the Sun.

these movements actually existed. They merely had to observe their consequences: the rising and setting of the Sun; the waxing and waning of the Moon; and the annual cycle of the Sun's position in the sky. The early history of astronomy was much concerned with the definition and measurement of these three 'natural' units of time. Monuments like Stonehenge, whatever other purposes they served, were designed partly to enable the length of the year to be fixed with precision.

In ancient civilizations, the *day* was defined by the movement of the Sun, but this was used in different ways. The Egyptian day started at dawn. The Babylonian and the Islamic days began at dusk. The Chinese day, and the Roman day, like the later Christian day, began at midnight.

In the days before artificial light, the Moon played a greater part in people's sense of time; and the month – the length of the Moon's cycle of changes – was another 'natural' subdivision of time. Unfortunately, it wasn't a round number of days, nor did it fit neatly into a year of 365 days; so some months had to have extra days, or there had to be extra days that didn't belong to any month.

ARTIFICIAL TIME   The day, the month, and the year provide an obvious, and one might say, inevitable, framework for any system of time recording; but they are of limited use in operating a sophisticated civilization. The day, in particular, is useless as a basis for arranging meetings, or organizing work schedules. Nature didn't provide a natural unit for this purpose, so humanity had to make one up. This was the origin of the *hour*. Being arbitrary, the hour found different definitions in different societies. Most ancient civilizations divided a 'day' – measured from sunrise to sunset – into a fixed number of hours, with the result that an hour was longer in summer than in winter. This was no use to astronomers, and in the second century BC the Greek astronomer Hipparchus introduced what was essentially the hour we use today. He defined his day as $\frac{1}{12}$th of the interval between sunrise and sunset at the spring and autumn equinoxes, when the periods of darkness and light are of equal length. This *equinoctial hour* thenceforth became the definition used by astronomers; and it was a measure that could be used to calibrate water clocks and hourglasses. But in everyday life, the practice of dividing up the hours of daylight into equal intervals continued largely unchanged for another 1,000 years. It wasn't until mechanical clocks made their appearance in the fourteenth century that Hipparchus' hour achieved general acceptance throughout Europe.

Minutes and seconds are arbitrary units. They too were introduced by Hipparchus, who hit upon the idea of subdividing hours by the Babylonian unit of 60, and then by 60 again. This gave a very neat result, in that the smallest unit of time that resulted – the second – was about the length of a resting heartbeat.

One other artificial, or contrived, measure of time has been a feature of calendars for over 3,000 years: the *week*. This has proved useful for both religious and civil purposes. By far the most widely used definition of a week has been 7 days, because 4 weeks of 7 days fit fairly neatly into 1 cycle of the moon. But other weeks have been used. The Roman week contained 8 days.

DAYS AND YEARS Astronomers discovered early on that there were problems in reconciling the natural units of the day and the year. The year, as measured by the changing height of the Sun above the horizon, or the reappearance of the constellations, does not equal a round number of days. And the day itself can be defined in various ways.

Our everyday definition of a day is called the *solar day*. This is the time it takes the Sun to return to the same position in relation to any given point on the Earth's surface. This is the day we divide into 24 hours. But the Earth actually completes 1 revolution on its axis in 23 hours 56 minutes 4 seconds. In other words, the Earth revolves 366 times every 365 days. Astronomers call this shorter period of the Earth's revolution on its axis – which is measured in relation to the stars, not the Sun – the *sidereal day*. The reason for the difference is that every day the Earth completes $\frac{1}{365}$th of its annual journey around the Sun. As a result, it has to make $\frac{366}{365}$ths of a revolution for any given point on its surface to get back into the same position in relation to the Sun as it was the day before.

A year – the time it takes the Earth to complete its annual journey around the Sun – is not exactly 365 solar days. Roughly speaking, a year is $365\frac{1}{4}$ days; which is why we need an extra day (29 February) every 4 years, to keep the calendar in line. And even that doesn't give us quite the result we need, because a year is actually only 365.242 days. So 3 times in every 4 centuries – for example, in 2100, 2200, and 2300 – we don't have a leap year. But if the year is divisible by 400, e.g. 2400, we do.

PRECESSION   The Earth spins like a top. But it doesn't just spin like a top, it also *sleeps* like a top. As a top spins, its axis describes an inverted cone, in a movement known as 'sleeping'. The Earth does the same. And, as it does so, an imaginary line, drawn through its axis, up into the northern sky, sweeps out a circle among the stars. This motion – called *precession* – was discovered by the Greek astronomer Hipparchus in the second century BC.

Precession is a slow process. It takes 26,000 years for the Earth's axis to complete one turn. One consequence of precession is that the Pole Star, which has for so long been a friend to travellers in the northern hemisphere, will in time become less useful, as the Earth's axis comes to point to a different part of the northern sky. The explanation of precession is to be found in two facts: that the Earth's axis is tilted at an angle of 23 degrees to the plane of its orbit, and that the Earth bulges slightly at the equator. The pull of gravity from the Sun – and even more, the Moon – on this bulge sets up the wobble we call precession.

This comparison of precession with the behaviour of a top, which is the way in which it is usually described, is not a very accurate one. A top really does behave as if its axis moved around the surface of an inverted cone. But the Earth's axis describes *two* cones that meet at its centre. An extension of the Earth's axis, projected into the southern sky, also describes a 26,000-year circle against the background of the stars.

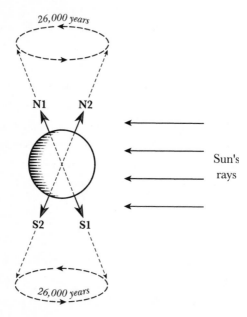

Figure 3. Precession: the Earth's Wobble
The Earth performs a wobble around its centre, once every 26,000 years.
N1, S1 are the present positions of the poles in December.
N2, S2 show December positions 13,000 years from now.

MEASURING THE EARTH After Aristotle, the centre of Greek scientific thought moved to Alexandria, in Egypt. It was there, around 200 BC, that one of the first great feats of practical astronomy – the measurement of the circumference of the Earth – was performed. The person responsible was Eratosthenes of Cyrene, the head of the city's library, which at that time was the most illustrious centre of learning in the Western world.

The calculation he performed was beautiful in its simplicity. He had heard that at noon at the summer solstice, 21 June, the sun at Syene, 800 kilometres/500 miles south of Alexandria, shone directly down a well. Knowing that the sun at Alexandria on the same date was not directly overhead, he decided to measure the difference in the sun's angle at the two locations. He placed a stick in the sand, and measured the length of its shadow at noon. Using elementary trigonometry, he deduced that the sun's rays at Alexandria fell at an angle of 7.2 degrees to the vertical. Assuming the two places were 800 kilometres/500 miles apart, he calculated that the circumference of the Earth – over the poles – was (360 ÷ 7.2) × 800 = 40,000 kilometres/25,000 miles, which was less than 1 per cent out.

Eratosthenes didn't *know* the two places were 800 kilometres/500 miles apart. He guessed the distance from the time it took a camel train to make the journey. And Syene wasn't *exactly* south of Alexandria. Nor would the sun there have shone *precisely* straight down a well on 21 June. So he was a bit lucky in his answer. But his mthod was correct, and the intellectual daring he displayed in tackling such a colossal problem with such simple equipment ensured his undying fame.

THE EARTH'S VITAL STATISTICS

Equatorial diameter: 12,760 km/7,926 miles

Equatorial circumference: 40,100 km/24,900 miles

Polar diameter: 12,720 km/7,900 miles

Polar circumference: 39,960 km/24,818 miles

Mass: 5.88×10^{21} tons[*]

Mean density:[**] 5.5

Escape velocity (at the surface): 11.2 km/6.96 miles per second

Distance from the Sun: (maximum) 152 million km/94.5 million miles

(minimum) 147 million km/91.4 million miles

(average) 149.5 million km/92.9 million miles

Speed of rotation at the equator: 1,675 km/1,040 miles per hour

(15 degrees per hour)

Mean orbital velocity around the Sun: 29.8 km/18.5 miles per second

107,200 km/66,600 miles per hour

Orbital period (1 tropical or solar year): 365.242 days

Mean sidereal day:[***] 23 hours 56 minutes 4 seconds

Mean solar day:[****] 24 hours

Inclination of the Earth's axis: 23 degrees, 27 minutes

Length of a degree of longitude at the equator: 111.4 km/69.2 miles

Length of a degree of latitude at the equator: 110.6 km/68.7 miles

Length of a degree of latitude at the poles: 111.7 km/69.4 miles

Surface area: (land) 29%

(ocean) 71%

[*]10^{21} means 1 followed by 21 zeros. The mass of the Earth is 5,880 million, million, million tons.

[**] The mean density of water is 1.

[***] The time taken by the Earth to complete one revolution on its axis.

[****] The interval between one midnight and the next.

COLUMBUS' MISCALCULATION   Seventeen centuries after Eratosthenes arrived at a good estimate of the Earth's circumference, a 40-year-old adventurer of Genoese origin named Cristobal Colon (our Christopher Columbus) sailed west across the Atlantic. He had been encouraged by a less accurate estimate. By the late fifteenth century, the belief in a spherical Earth was almost universal among educated people in Europe. Columbus got the idea of attempting to reach the east by sailing west from the Italian mapmaker, Toscanelli. His belief was that he faced a journey of only 6,300 kilometres/3,900 miles. He left the Canary Islands in September 1492, convinced that all he had to do was to sail west along the 38th Parallel, and he was bound to reach 'The Indies', as East Asia was then known. Unfortunately – or fortunately, as it turned out – there was a continent in the way. One can only wonder whether he would have bothered to get out of bed if he had been working with Eratosthenes' more accurate calculation, which would have promised him a voyage of 22,000 kilometres/14,000 miles.

ANCIENT CHINESE CIVILIZATION   While the Greeks were working out the ideas that would later provide the launch pad for the development of modern science, a great civilization was in full flower in China, 10,000 kilometres/6,000 miles to the east. The Greeks knew little of it. Had they known more, their sense of their own cleverness might have suffered a shock. In astronomy, in literature, painting and pottery manufacture, in military technology and in public administration, the Chinese achievement was the equal of anything the Greeks could show. In iron working, in civil engineering, and in agriculture, they were far ahead. In areas such as silk manufacture and calligraphy, they had perfected arts and crafts of which their Western contemporaries had no conception.

If Greek philosophers of the first century BC could have been transported to China, they would have been amazed at the level of technology: ploughs with integral iron shares; deep drilling for brine and natural gas; the manufacture of steel from cast iron; mass-produced cross-bows; and harnesses that enabled horses to haul extraordinary loads. However, they might have been puzzled by the absence of the sort of scientific speculation that was meat and drink to them. And they would certainly have been surprised at the lack of progress in some subjects – for example, geometry – that were central to their own thinking. But they would have been left in no doubt that they were in the presence of a great civilization.

A GREAT CHINESE SCIENTIST Zhang Heng (or Chang Heng) was an example of the type of scientist that ancient China was capable of producing. Born in Nanyang, central China, in 78 AD, he was one of those impossibly gifted geniuses who make ordinary mortals feel as if they belong to a different species. The range of his talents puts one in mind of Leonardo da Vinci; except that, as a scientist, he was clearly Leonardo's superior. He was one of the four great painters of his age, and he produced 20 famous works of literature. But he was first and foremost an astronomer. He was astronomer royal under the East Han Dynasty, in the second century AD. He produced one of the world's first great star maps, rivalled only by the one that had been created by Hipparchus, unknown to Zhang, in 129 BC. On this map he plotted accurate positions for 2,500 bright stars, giving names to 320 of them. He estimated that the night sky, only part of which was visible from China, contained 11,500 stars. This was a bit of an exaggeration, even for a keen-sighted observer, but it was not a bad estimate. He explained eclipses of the Moon, correctly, as being caused by the Moon's passing through the Earth's shadow. And he pictured the Earth as a small sphere suspended in space, surrounded by a vast, and far-distant, spherical sky. Zhang Heng was also a gifted mathe-matician, and he improved previous estimates of the value of *pi* (the ratio of the circumference of a circle to its diameter) from a rough-and-ready value of 3 to 3.162, which compares well with the figure of 3.142 accepted today.

Zhang Heng's most famous piece of work was the earth-quake detector he perfected in 132 AD, 1,700 years before the first European seismograph. Zhang amazed the imperial court with this device, which could detect distant earthquakes that no one near it could feel. It took the form of a bronze vase, to which were affixed a number of bronze dragons' heads, each holding a bronze ball in its mouth, and around the foot of which were a number of bronze toads with open mouths. If the machine detected an earth tremor, a ball was automatically released, falling into the mouth of one of the toads. The posi-tion of the particular toad involved indicated the direction from which the tremor had arrived. On one famous occasion, a ball dropped without any perceptible tremor being observed. It was not until several days later that a messenger arrived with news of an earthquake in Kansu, 600 kilometres/400 miles away from the court, in the direction indicated by the machine.

Despite the brilliance of Zhang Heng's creation, it is wrong to credit him with the invention of the seismograph. His machine was an *earthquake detector*. It recorded earthquakes, but it didn't measure them.

CALCULATING *PI* *Pi,* the number Zhang calculated as 3.162, cannot be expressed precisely in numerical terms, either as a common fraction or as a decimal fraction. No matter how many digits are used, the answer can only be an approximation. A value of 3.1416 is as accurate a figure as most people will ever need for practical purposes. Before computers, the greatest number of decimal places to which anyone had been able to take the calculation without making a mistake was 528. However, in 2002, a Japanese team succeeded in calculating *pi* to 1.24 *trillion* decimal places. But that's still only an approximation.

ISLAMIC SCIENCE Whatever may have been the achievements under the centuries of Roman imperial rule, the advancement of science was not one of them. Greek science was remembered by the Romans, but little was added to it. When the empire disintegrated, city life was largely abandoned, and scientific knowledge was lost. With the rise of Christianity, intellectual activity became centred mainly on theology rather than 'pagan' science. Had it not been for the rise of another empire further to the east, most of this ancient knowledge would have been lost, and the history of science would have been very different.

In the year 610 AD, a 40-year-old merchant named Muhammad, who lived in Mecca, in the Arabian peninsula, experienced a series of revelations, which became the basis of the *Qur'an*, the sacred book of a new religion. Within 20 years, this religion, Islam, had spread throughout most of Arabia. Under a succession of military and religious leaders known as caliphs, Islam exploded across the Middle East and North Africa, conquering two older empires, Byzantium and Persia. By the year 750, little more than a century after Muhammad's death, the empire of Islam stretched from the Bay of Biscay to the mountains of Afghanistan.

An important element of Islamic teaching was the obligation to pursue knowledge. This imperative, and the wealth generated by the trade of a vast empire, created an environment in which science flourished. Under a succession of rulers sympathetic to science, a great gathering together of learning took place. From Toledo in the west to Isfahan in the east, scholars worked on translations of ancient texts: not just from Greek, but also from Sanskrit, Syriac, and Pahlavi. At the same time, trading contacts with China and India introduced ideas and mathematical techniques that had been unknown to the Greeks. It can be said with confidence that Islamic scholars of this time were in possession of a greater body of scientific knowledge than had ever before been assembled in the history of the world.

DISCOVERIES IN BAGHDAD   Throughout the empire, trade created wealth, and wealth created population. New cities arose, like Cordoba in Spain, which grew to house half a million people. Many of these cities became centres of learning. But nowhere in Islam did the light of science shine more brightly than in Baghdad, on the banks of the river Tigris in Mesopotamia. Here, on the northern borders of ancient Babylonia, where the science of astronomy had been born 3,000 years earlier, a great city arose, to which scholars came from far and wide. Over the course of 3 centuries, its size and prosperity increased, until by the year 1000 its population reached 1.5 million. In this environment, science could not help but flourish. In the early ninth century, when the Emperor Charlemagne could barely write his name, the caliph of Baghdad, Harun al-Rashid, and his scholars were exploring and extending the boundaries of mathematics, astronomy, geography, and medicine. In 830, al-Rashid's son, al-Ma'mun, founded the House of Wisdom, where scholars worked on translations of works by Aristotle, Archimedes, Ptolemy, and others that had been recovered from the far corners of the empire.

There is a story about al-Ma'mun that sums up the attitude of early Islamic scholars towards the Greeks. It is told that, in a vision, he saw a man reclining on a couch, and asked 'Who are you?' When the man replied, 'Aristotle', the delighted Caliph initiated a long discussion on morality, law, and faith. This open-mindedness meant that, for 600 years, Baghdad and other Islamic cities remained treasure-houses of scientific knowledge. This is how it was preserved until the day when Europe woke from its intellectual slumbers and embarked on its own voyage of scientific discovery.

THE ARABIC SYSTEM OF NUMERALS Of all the gifts that Islam gave to modern science, the most important was the system of numbers we call *Arabic numerals*. If we were to give credit where it is ultimately due, we would call them *Indian numerals,* since they originated in India in the first millennium BC. They were put into regular form by the Hindu astronomer Aryabhata, who was born in Kusumapura, near present-day Patna, in 476. He outlined the system in a book on astronomy and mathematics called the *Aryabhatiya,* which was written in Sanskrit verse couplets. His book was not published in Europe until 1874. But the system of numbers it described had long before found its way into Arabic in the ninth-century writings of the mathematician al-Khwarismi, who was born in what is now Uzbekistan. He was one of the men of science who found a patron in the caliph al-Ma'mun; and it was through him that Arabic numerals found their way into Europe, in time for the revolution initiated by Galileo and Kepler.

It is not necessary to have an efficient system of numbers to do rapid calculations. A skilled operator with an abacus can perform some very impressive operations. But the kind of paper-and-pencil calculations involved in the problems that Newton faced would have been impossible without Arabic numerals, decimal-place values, and the concept of zero, all of which we owe to India, by way of Islam. Anyone with doubts on the matter should try multiplying MDXXLIV by LIX, or consider how they would use Roman numerals to handle calculus.

The use of Arabic numerals in Europe – at first for book-keeping, and only later to perform scientific calculations – was initiated by the Italian mathematician Leonardo Fibonacci, who described them in his *Book of Calculation*, and who demonstrated their scientific use in his *Book of Squares,* which was written in 1225.

THE WORK OF COPERNICUS   In the autumn of 1491, as Columbus was planning his voyage, an 18-year-old student began his university studies in the Polish capital city of Krakow. His name was Mikolaj Kopernik, which he later changed to its Latin equivalent, Nicolaus Copernicus. He was destined to bring about as great a change in people's image of the regions beyond the Earth as Columbus did of their image of the Earth itself.

Copernicus' parents died before he was 10, and he was brought up in the house of his uncle, who was a bishop. When he was 22, his uncle secured him a lifetime appointment as a canon of Frauenburg Cathedral. It was not an onerous appointment, so he was able to continue his studies. Attracted by the flow of ideas coming out of Italy, he enrolled at the university of Bologna, where he assisted the astronomer Domenico Maria da Novarra, and afterwards at the universities of Ferrara and Padua.

Although he continued his study of church law, astronomy was his first love. When he began his studies, the accepted model of the heavens was still the one constructed 1,300 years earlier by the Greek astronomer, Ptolemy of Alexandria. Ptolemy's universe was an Earth-centred one, in which the Moon, the Sun, the planets, and the stars all revolved around a central, stationary Earth.

Ptolemy was born around 100 AD and died around 170 AD. Although he wrote in Greek, he may have been Egyptian. His claim to fame is his book, *The Mathematical Collection*, in which he summarized the astronomical knowledge of his age, and which we know by its Arabic title of *Almagest* ('The Great Book'). His Earth-centred universe was the model accepted by most astronomers in Copernicus' day, and it was the one endorsed by the Church. But it was a model that had become less and less reliable as a guide to the movements of the planets. And the intricate calculations that were needed to make sense of Ptolemy's system raised questions in the minds of some astronomers, who felt that the movements of the heavens ought to display a simpler and more elegant pattern.

These were questions that naturally occurred to a young man of restless intellect like Copernicus. When he finished his studies in Italy, he returned to his uncle's estate, where he continued to mull over the problem of planetary movements. He was not an astronomer, in the sense of someone who studies the heavens by observing them. He was a natural philosopher: someone who thought deeply about the natural world, but who got the raw material of his thinking from reading and reflection, rather than experiment and observation. It was his reading of Greek texts that pre-dated Ptolemy that gave him the idea of a Sun-centred universe. When his uncle died in 1512, Copernicus returned to Frauenburg. Within two years he was circulating among his friends a document which he called his *Short Commentary*, containing his thoughts on the limitations of the Ptolemaic system.

Before Ptolemy wrote the *Almagest,* there had been other Greeks who had subscribed to a model in which the Earth and the planets revolved around a central Sun. But they had not worked their ideas up into a fully fledged system as Ptolemy had, and their names had mostly been forgotten. Dissatisfied with what he considered to be a defective model, Copernicus began to play with the idea of a Sun-centred system, even though he was so conditioned by the conventional wisdom that he thought it must be a ridiculous idea. But the more he tested it against the observed behaviour of the planets, the more sense it seemed to make.

There was one phenomenon, the 'retrograde' motion of the planets Mars, Jupiter and Saturn, that required an especially convoluted explanation in Ptolemy's model. This was the way these planets seemed, from time to time, to stop in their journeys against the background of the stars, and to go backwards for a while, before resuming their forward motion. Copernicus realized that this was exactly what one would expect to see if these planets were revolving around the Sun, and the Earth, on its shorter journey, overtook them on the inside.

Pleased as he was with the simplicity of his model, he feared ridicule if he were to publish such a revolutionary proposal. He would probably never have gone public, had a young mathematician not travelled from Germany to seek him out in 1539.

Copernicus' admirer was a professor of mathematics in the university of Wittenburg, named Rheticus, who had read his *Short Commentary* when it was circulated a quarter of a century before. Encouraged by this new disciple's enthusiasm, and discovering that the Pope himself wished to see his ideas in print, Copernicus agreed to let Rheticus supervize their publication, in a book entitled *On the Revolutions of the Heavenly Globes,* which was published in 1543. If legend is to be believed, the first copies arrived from the printer on the day Copernicus died.

THE SUN Our nearest star, the Sun, seems special to us, but in cosmic terms it is pretty ordinary. It is an average sort of star, neither particularly big nor particularly small. It is a globe of hot gases, mostly hydrogen (about 75 per cent) and helium (about 25 per cent), and it is just one of several hundred billion stars swirling around the Galaxy – our disc-shaped local universe. Situated well away from the centre of the disc, it is hurtling along at 241 kilometres/150 miles a second; even so, it takes more than 200 million years to go around the Galaxy once. The

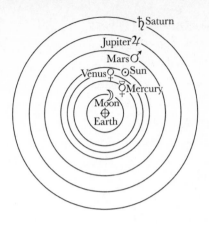

Figure 4. The Solar System According to Ptolemy (c 140)

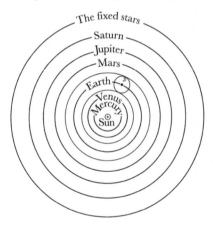

Figure 5. The Solar System According to Copernicus (1543)

Sun's average distance from the Earth is 150 million kilometres/ 93 million miles; but because the path of the Earth's orbit around the Sun is an ellipse, not a circle, the actual distance varies by about 1 per cent on either side of this figure. The Sun's light takes 8 minutes and 20 seconds to reach the Earth.

THE SUN'S VITAL STATISTICS

Diameter: 1,400,000 km/864,000 miles (100 times that of the earth)

Volume: 1,300,000 times that of the Earth

Mass: 330,000 times that of the Earth

Average density:* 1.4

Period of rotation: 25 days

Surface temperature (measured): 5,500°C/10,000°F

Core temperature (calculated): 15,000,000°C/27,000,000°F

* The density of water is 1.

The Sun is accompanied on its journey by its attendant planets, which make up...

THE SUN'S FAMILY

	Diameter (million)		Mean distance from the Sun (million)		Orbital period
	(km)	(miles)	(km)	(miles)	
					(days)
Mercury	4,900	3,000	58	36	88
Venus	12,100	7,500	108	67	225
Earth	12,800	8,000	150	93	365
					(years)
Mars	6,800	4,200	228	142	2
Jupiter	143,000	88,800	779	484	12
Saturn	120,600	74,900	1,433	891	29
Uranus	51,200	31,800	2,873	1,785	84
Neptune	49,600	30,800	4,495	2,793	165
Pluto	2,400	1,500	5,870	3,647	248

THE ASTEROIDS   The planets are the senior members of the Sun's family; but they have a swarm of little brothers and sisters – the asteroids – that race around like skateboarders on a quiet stretch of road: usually managing to stay clear of traffic, but occasionally failing to get out of the way in time.

The asteroids are mostly located in a belt between the orbits of Mars and Jupiter, but some have orbits uncomfortably close to the Earth's. Some are lumps of metal, or rock and metal (mainly iron), but most are just chunks of rock. They are thought to be remnants of material that failed to consolidate into planets. Ceres, the biggest, was discovered in 1801, and is about 1,000 kilometres/600 miles across – about a third of the diameter of the Moon. There are 30 or so with diameters of more than 150 kilometres/100 miles.

The answer to the question, 'How many asteroids are there?' depends on how big a lump you believe constitutes an asteroid. If you're not fussy, the answer is 'millions'. More than 50,000 have been named or numbered, and even the smallest would be capable of giving the Earth a nasty smack, while an asteroid 1.5 kilometre/1 mile wide, entering the Earth's atmosphere at anything up to 80,000 kilometres/50,000 miles an hour, would cause quite a bump – and has done, more than once, as the geological record shows.

It isn't a question of whether a sizeable asteroid will cross the Earth's path, merely a question of when. Fortunately, it doesn't happen often, and when it does, we won't hear it coming. We won't even see it, unless we happen to be look-ing in its direction just before it hits us, so there is no point in worrying.

THE KUIPER BELT The asteroid belt between Mars and Jupiter has been a subject of study by astronomers for 200 years; but it is only in the last few decades that we have become aware of a similar group of objects hurtling around in the outer reaches of the solar system. This is the Kuiper Belt, a region just outside the orbit of Neptune, which is populated by millions of what are thought to be comet-like lumps of ice, rather than the rock-like asteroids of the inner belt. The largest so far discovered is about 1,300 kilometres/800 miles in diameter. Some astronomers think that the 'planet' Pluto, which has a very eccentric orbit, and is mainly composed of rock and ice, is really one of these Kuiper Belt objects; but, for the moment at least, the International Astronomical Union still classifies it as a planet.

METEORITES *Question:* When is an asteroid not an asteroid? *Answer:* When it's a meteorite. While a lump of rock or iron is cruising along in its orbit around the Sun, we call it an asteroid. If we find one after it has fallen, we call it a meteorite.

Small meteorites are dramatically slowed down by the Earth's atmosphere, and much of their mass is vaporized in the process. But quite a few reach the ground every year, and the Earth's surface is littered with them. Fortunately, the Earth's surface is mostly water and empty land, so they seldom hit anything that matters: there is no record of anyone having been killed or injured by one – so far. But their potential for doing damage is spectacular. A moderate-sized meteorite, not much bigger than a house, travelling at 800 kilometres/500 miles a minute, packs a similar punch to the bomb that destroyed Hiroshima. Impacts on this scale occur on average about once every 5,000 years.

Occasionally, the Earth meets up with a big one. There is a buried crater 200 kilometres/120 miles wide at Chicxulub, on the Yucatán peninsular in Mexico, that was gouged out 65 million years ago by a meteorite about 10 kilometres/6 miles wide. The cloud of debris that this flung around the world probably played a part in the demise of the dinosaurs.

For people of a nervous disposition, it helps to know that there is less chance of being hit by a meteorite on a morning walk than on an evening one.

COMETS There is another very different class of objects travelling around the Sun: the comets. They are not very big. The largest are only a few miles across, and most are much smaller. They are mostly lumps of frozen water, with an admixture of frozen methane, carbon dioxide, etc. But this doesn't make them harmless. What matters is the combination of velocity and mass. Comets, which mostly come from the far reaches of the solar system, travel at very impressive speeds; if the Earth met up with one, we would certainly know it. Their orbits are more markedly elliptical than the orbits of the planets. They sweep in close to the Sun, and then swing out again. Some return every few years, but others travel so far out in space that they disappear for centuries at a time.

In the days when astrology was the end, and astronomy merely the means to that end, comets were regarded with superstitious awe. Appearing from nowhere, with their bright heads and their streaming tails, and moving slowly across the sky over several weeks, they really did seem like portents of great events.

It is only when a comet nears the Sun that it assumes the appearance that artists like to portray. Under the influence of the Sun's heat, some of its substance evaporates, and the *solar wind* – a stream of charged particles emitted by the Sun – pushes this evaporated material away from the main body of the comet. This is the origin of the comet's 'tail', which always points away from the Sun: this trails when the comet is approaching the Sun, but actually points forward when it is moving away.

Most *short-period* comets (those with periods of less than 200 years) are thought to originate in a region beyond Neptune called the Kuiper Belt, and have been diverted by the gravitational attraction of the major planets. *Long-period* comets are thought to originate in an even more distant region known as the Oort Cloud. The Oort Cloud, far beyond the orbit of Pluto, contains millions; and some of them are substantial. Every now and again, one gets disturbed by the gravitational attraction of a planet, or perhaps a nearby star, and swings Sun-wards. Hopefully, there isn't one with our name on it.

METEORS AND METEOR SHOWERS   Meteors, or 'shooting stars' are different from meteorites. They are tiny fragments of rock or dust that are completely burned up by the friction of their passage through the upper atmosphere. They appear as bright streaks across the sky, lasting for a second or less. Some travel in streams in the wake of comets. When the Earth crosses their path, we are treated to a meteor shower. A meteor shower is not as spectacular as it sounds, but some of the more reliable ones can serve up several meteors a minute. The regular showers take their name from the constellation they appear to come from. The following are the best known:

Name of shower	Date (and peak)
Quadrantids	1–5 January (4 January)
Lyrids	16–25 April (22 April)
Perseids	23 July–22 August (12 August)
Orionids	2 October–4 November (21 October)
Leonids	14 November–21 November (17 November)
Geminids	6–19 December (13 December)

WHAT MAKES SCIENCE POSSIBLE Some things in history are so familiar that we never stop to ask ourselves why they happened when and where they did, rather than at some other time and in some other place. The history of science is one of these things we take for granted. But if we stop to think about it, there is a riddle at its heart.

Greek science was unique in the ancient world. The Chinese created a great civilization; and their technology was in many ways hundreds of years in advance of the rest of the world. Ships, weapons and agriculture, roads, bridges and locks, paper and printing: the list of their technical innovations goes on and on. Nothing like it had been seen before; and nothing comparable would be seen until the agricultural and industrial revolutions in early modern Europe. But the more one tries to make out a similar case for Chinese science, the more clear it becomes that, as scientists, the Chinese were just not in the same league. Even the historian Joseph Needham, who opened the world's eyes to Chinese achievements in technology, confessed himself puzzled by their failure to make comparable progress in science.

Now consider the Romans. They also created a great civilization. They too had magnificent technology. They didn't get round to inventing paper or gunpowder, but when it came to roads and bridges, aqueducts and steam baths, communications, and the administration of a great empire, they were easily the equal of the Chinese. But hundreds of years of Roman civilization produced next to nothing worthy of being called scientific advance. They had Greek slaves, and access to the whole of Greek science, and could have built on that, had they felt the urge. Yet, when the classics of Greek science came to be translated into Latin in the sixteenth century, it was to Arabic versions that the translators had to turn for the majority of their texts.

Science, as we understand it, has only happened twice in the history of the world. And between the twilight of the Greek world and the dawn of the modern scientific age there was an interval of a millennium and a half, during which little of any consequence was added to the world's stock of scientific knowledge. Why should this be? It can't be in the genes. The Greeks were no cleverer than the Romans or the Chinese, nor were the inhabitants of Europe cleverer than the Aztecs, or the people of Great Zimbabwe. Perhaps the economy is the key: science can only prosper in societies rich enough to enable a lot of people to sit around thinking and talking. But wealth, and leisure, and urban living, can't be the whole story, or ancient Rome and classical China would have been scientific powerhouses. The explanation must be cultural as well as economic. Some societies are organized in ways – and develop habits of thought – that make science possible; others, equally prosperous, have social and political arrangements, codes of belief, and ways of thinking that stifle science. Societies with an exaggerated respect for the past cannot generate the challenging attitude to accepted ideas that produces new understanding. Societies in which priests have power are liable to imprison or otherwise suppress those who threaten their monopoly of explanation. Where free speech and free thought are constrained, minds, as well as bodies, rot in chains.

Science is a plant that needs favourable conditions. It cannot grow in a wilderness, nor does it thrive in darkened rooms. It grows best in towns (including those town-like places called universities), nurtured by people with the means and the leisure to care for it. It needs light and air, and a fertile soil. In Europe in the sixteenth and seventeenth centuries, the conditions were right, and science burst into luxuriant growth.

SCIENCE AND TECHNOLOGY   The expansion of scientific knowledge that occurred in the early seventeenth century is often attributed to the rediscovery of ancient learning that sparked the phenomenon of the Renaissance. But the more one examines this explanation, the less adequate it is. Between Archimedes and Eratosthenes there were 400 years of speculation about the natural world, by some of the best minds that science has known: but they can hardly be said to have accumulated a vast store of knowledge about how the natural world works. Had science between 1600 and 2000 moved at the same pace, starting from the knowledge the Greeks possessed, and using the tools the Greeks used, we would not have added much to the stock of knowledge we inherited from them. The rediscovery of ancient learning certainly provided a launch pad. But it needed something from outside science to propel science into orbit: something neither the Greeks, nor the Arabs, nor the Chinese had. That something was the right technology.

Technology is defined in many dictionaries as *applied science*, but it is no more meaningful to define technology as applied science than it is to define a hen as an applied egg. Hens do indeed come from eggs, but eggs also come from hens. It is true that much new technology has come from the application of scientific discoveries, but it is equally true that scientific discoveries have often been the result of the exploitation of new technology. Science and technology are simply two different responses to the forces of nature. Science is humanity's attempt to *explain* them. Technology is humanity's attempt to *exploit* them. And progress in either can be the source of progress in the other.

THE PRINTING REVOLUTION Of all the technical innovations that ushered in the age of modern science, one stands out: the invention of printing. Or rather, the invention of the printing *press*, using movable type. Printing itself wasn't new. The Chinese had been printing books for hundreds of years. What was new was printing as a form of mass production. In the fifteenth century, Chinese science and technology were the equal of anything that Europe possessed, and Chinese book production was superior. Europe had beautiful manuscripts. But they were handcrafted products, written with the feathers of birds, and many of them were still written on the skins of sheep. The Chinese had beautiful manuscripts *and* beautiful books, printed on the paper they had invented. But their books were costly to produce. They were printed from blocks, and virtually every page required its own specially carved block. Once the blocks had been prepared, multiple copies could be produced, but the carving of the blocks with intricate Chinese script, and the subsequent hand-inking, were slow processes.

In late fifteenth-century Europe, a number of previous inventions were brought together to create what we now call printing. It was this, more than any other single factor, that ensured that Europe would be the birthplace of modern science. It was one of the most significant events in history. If the Scientific Revolution, the Agricultural Revolution, and the Industrial Revolution merit capital letters, so too does the Printing Revolution.

ALPHABETS AND MOVABLE TYPE One innovation that made the Printing Revolution possible – the alphabet – was already 2,500 years old. This clever idea, which makes it possible to express any thought, and convey any information, using just 20 or so simple symbols, originated in the eastern Mediterranean around 1000 BC. It was adopted by the Greeks and then by the Romans. By the year 1400 some version of it was the written medium of every language in Europe.

Another key innovation was the use of movable type, made from cast metal. This was not a uniquely European invention. The Koreans had been using movable type for 200 years, and they had passed the idea on to the Chinese. But it never made much headway in East Asia, because of the cost of the huge number of symbols required. It was much easier for Europeans to exploit movable type, because the stock of letters they needed, and the capital investment involved, was much smaller than it would have been with scripts like Chinese or Japanese. The time and cost involved in typesetting were also less.

Given the alphabet, and movable type, it needed just one other innovation – the printing press – to bring about the revolution. It is not surprising that the revolution took place in Europe, where people had been using the screw press to crush olives and grapes since its invention by the Romans.

All these contributing inventions – paper, smooth-running ink, the alphabet, movable type and the screw press – were finally brought together in the German city of Mainz. The time was the 1450s, and the man responsible was a skilled metal-worker named Johannes Gutenberg.

THE MARKET FOR BOOKS   Gutenberg was not a scientist, and he was not concerned with the advancement of scientific learning, or any other kind of learning. He was an entrepreneur, who saw an opportunity and set out to exploit it.

In fifteenth-century Europe there was an unsatisfied demand for the written word. Increased prosperity had created a new class of people with money to spend on entertainment. Anyone who could supply them with stories to read was assured of a ready market. Prosperity had also created a new class of leisured scholars. Between 1400 and 1500 the number of universities in Europe increased from 20 to 70. Churches offered a mouth-watering market for bibles and prayer books. The opportunity was plain to see, and Gutenberg was only one of many engaged in the race to develop a new method of printing. But he was first across the line. By 1439, he was already printing from a press with movable type, though none of this early work survives. His name now is inseparably associated with the Gutenberg Bible, which was printed simultaneously on six presses and published in 1456.

Gutenberg's invention did not make him rich. Had modern patent law existed in the fifteenth century, he could have been the Bill Gates of his day. But he received no royalties from the hundreds of rival printers who exploited his invention. And having fallen out with the partner who financed his own operation, he even lost control of that. But he remained in favour with his archbishop, who ennobled him, and ensured that he lived out his latter years in comfortable security.

Once Gutenberg had shown the way, the books poured out. Printers took their skills to other countries. Italy's first press was established in 1464, Paris had presses by 1470, and London by 1476. By 1500, every European country except Russia had joined in, and 8 million books in 40,000 editions had been produced and sold. This explosion of supply in turn caused a further explosion of demand. No one who was in a position to learn to read could any longer afford not to do so.

MICROSCOPES AND TELESCOPES If printing made modern science possible, the technologies that brought it into existence were glass-making and lens-grinding. Lenses were not new. Spectacles had been in use in Europe and in China since the thirteenth century, but they were made of quartz, not glass. For reading glasses, quartz was better than nothing; but for scientific purposes it was useless. It was not until strong, clear glass, and the techniques for grinding it, were perfected in the sixteenth century, that scientific lenses became practicable. When they did, two powerful new tools became available: the microscope and the telescope. And it was the discoveries made with these tools that launched science on the upward path it has followed ever since.

The telescope and the microscope both appeared around 1600. The inventor of the compound microscope – two magnifying lenses used in combination – was a Dutchman, Zacharias Janssen, who produced his first instrument around 1590. It was another Dutchman, Anton van Leeuwenhoek, who showed the world what the single-lens microscope was capable of. The telescope was also a Dutch invention. The credit for this is usually given to a spectacle-maker named Hans Lippershey, who lived in Middelburg, the capital of the Dutch province of Zeeland. It is said that either he or his apprentice made the discovery accidentally in the autumn of 1608, as a result of looking through two lenses at a nearby church steeple. It was Lippershey who had the idea of enclosing the lenses in a tube. He sold his invention to the Dutch government, which recognized its military significance, and tried to keep it secret. But it was a secret that couldn't be kept for long. By the following spring, telescopes were being sold as novelties in the street markets of Paris; and in July of 1609, news of them reached the ears of a 45-year-old Italian mathematician named Galileo Galilei.

GALILEO'S EXPERIMENTS   Galileo was born in Pisa on 15 February 1564. His father was a mathematician who set his son to study medicine. But, after hearing a lecture on geometry, Galileo read Archimedes, and resolved to devote his life to mathematics.

Unlike his contemporaries, whose usual way of responding to a question was to ask, 'What did Aristotle say?', Galileo believed in experiment. His first significant discovery was made when he was 17, and studying medicine. During a service in Pisa Cathedral, he noticed a chandelier swinging above him. He observed that from time to time a draught would cause it to swing more widely and, apparently, more quickly. Using his pulse as a clock, he timed the swings; he discovered that no matter how wide the swing, the time taken remained the same. When he continued the experiment with a pendulum at home, he discovered that the duration of its swing, irrespective of its amplitude, depended solely on its length.

For the first 20 years of his adult life, Galileo was a professor of mathematics: first at the university of Pisa, and later in Padua. He developed an interest in mechanics, especially the motions of falling bodies. It was not the easiest of subjects to study in an age without mechanical clocks. Early on he established an important truth: that the rate of fall was the same for heavy objects as for light ones. This contradicted what Aristotle had said, and what had been taken for granted ever since. Rolling balls down a gently inclined plane, Galileo was able to slow down their rate of fall sufficiently to allow him to measure their speed, using either a water clock, or his pulse. This enabled him to make three more discoveries. The first of these was that the time taken for the balls to fall from a given height was unaffected by the angle at which the plane was inclined. The second was that their speed of fall was subject to a continuous acceleration (which again contradicted Aristotle, who had taught that the speed of fall was constant). Most important of all, he discovered that there was a precise relationship between the height of the fall and the duration of the fall of the balls. The distance travelled increased in proportion to the square of the time elapsed (e.g., a body fell four times as far in six seconds as it did in three). It was a discovery that would later play an important part in helping Isaac Newton formulate his Theory of Universal Gravitation.

Another discovery that Galileo made was that the trajectory of an object such as cannonball was determined by a combination of its free-fall speed and the forward force applied to it. This discovery had obvious application to the practice of gunnery, but it was also of significance for the study of astronomy. One of the most persistent objections that had been raised to Copernicus' theory of planetary motion was that, if the Earth were truly revolving on its axis, anything not nailed down, such as tables, chairs, and people, would be hurled into space by centrifugal force. Galileo was able to show that this would not happen if the effect of this centrifugal force were less than that of the force that caused an object to fall naturally to earth. His ability to refute this particular objection not only helped to persuade him of the validity of Copernicus' theory, but would later help to persuade others too.

GALILEO AND THE POPE When Galileo heard of Lippershey's telescope in 1609, he knew nothing of its construction, but within weeks he succeeded in making one to his own design, with a magnifying power of three. In doing so, he set in train a series of events that was to have a tremendous influence on the future course of his life, and on the history of science. Before another month had elapsed, he had an instrument with a magnifying power of ten. With the aid of this simple instrument, he was able to resolve the Milky Way into a multitude of individual stars. He watched the moons of Jupiter circling their parent planet – proof positive that not every heavenly body revolved about the Earth. He looked at sunspots (permanently damaging his eyes), and used them to measure the speed at which the Sun revolved on its axis.

As a result of his observations, especially of the satellites of Jupiter, he became convinced of the soundness of Copernicus' theory that the Earth and the planets revolved about the Sun. It was not a conviction that endeared him to the rulers of the Church. In 1632, believing that the pope, Urban VIII, would not take his opinions amiss, he published them in a book entitled *Dialogue on the Two Chief World Systems*. It was a masterpiece; and it came close to getting him burned at the stake.

Faced with the Reformation, the Catholic Church was less tolerant of dissent than it had been in Copernicus' day. Anyone in a Catholic country who dared to publish ideas that undermined the teaching of the scriptures, or the authority of the Church, could expect a sharp reaction. In the Church's eyes, Galileo had overstepped the mark, and his offence could not be ignored. To assert that the supposedly perfect Sun had spots was bad enough. To advocate Copernican theory was worse. But what really did for him was the way in which he presented his *Dialogue*. He put the anti-Copernican argument in the mouth of a character he called Simplicio, whom the Pope was persuaded was meant to be an insulting caricature of himself. Galileo, now 69 years of age, was arraigned before the Inquisition, and compelled, on pain of death, to recant his belief that the Earth went round the Sun. He spent his last 8 years under house arrest on his little country estate, studying uncontroversial topics. With 4 years left to live, his sun-damaged sight failed him, and his stargazing ended.

THE ASTRONOMER TYCHO BRAHE There was one man above all others whom Galileo would have wished to tell about what he saw on that night in 1609 when he pointed his first telescope at the sky. But that man had died just eight years before, without ever having heard of this new instrument. His name was Tyge (in Latin, Tycho) Brahe, and he was perhaps the greatest naked-eye observer the world has ever known.

Tycho was born in 1546, on the family estate at Knudstrup, in southern Sweden. At the age of 13 he enrolled in Copenhagen University, where he studied law and philosophy. But in 1560 he witnessed an eclipse of the Sun, which persuaded him to switch to mathematics and astronomy. As a student, he already displayed the arrogant and argumentative temperament he would display all his life, even when dealing with royal patrons. When he was 19, his temper led him into a duel in which part of his nose was sliced off; for the rest of his life he wore a metal nose-cap.

After further study in Leipzig, Rostock, and Augsburg, Tycho returned to the family estate, where his uncle gave him permission to build an observatory. In 1572, he observed a bright new star in the constellation Cassiopeia. It was the first 'nova', or new star, ever recorded in Europe, and ever afterwards it would be known as 'Tycho's star'. It shone brightly for a year and a half before fading away. For a time, it outshone the planet Venus.

Tycho announced his discovery in a book entitled *Concerning the New Star*, in which he provided proof that his star was more distant than the Moon: a fact that went against conventional belief about the distance of the stars. His own guess was that it was 5 billion kilometres/3 billion miles away from the Earth, which turned out to be an enormous *under*-estimate. But, by the standards of the time, it was a stupendously daring figure to suggest.

The discovery of this new star struck the scientific world like a thunderbolt. Science in the sixteenth century was still in thrall to Aristotle, who had taught that the heavens were perfect and unchanging. The proof that this was not so raised the question of what else he might have got wrong. No scientific question could ever again be confidently answered with the words, 'Aristotle says...'.

Tycho's star made him famous. In 1576, the Danish king, Frederick II, granted him the island of Hven, near Copenhagen. Here Tycho built the best-equipped observatory in Europe. During the next 20 years he assembled a catalogue giving accurate positions for 777 stars, containing an unprecedented series of precise observations of planetary movements. His plotting of the brightest stars, and the positions of the planets, was correct to within 1 minute of arc – $\frac{1}{30}$th of the width of the full moon.

In 1588, Frederick was succeeded by Christian IV, who was less tolerant of Tycho's behaviour. In 1597, he took back his island, and cancelled Tycho's pension. In 1599, Tycho found a new patron in the Emperor Rudolph II, who gave him a castle outside Prague. He moved most of his equipment from Hven to Prague, but he had only two more years to live, and his observing days were effectively over.

Apart from his star catalogue, Tycho made many contributions to astronomy. He was the first astronomer to correct his observations to allow for atmospheric refraction; and his measurement of the length of the year was correct to a second.

KEPLER AND HIS LAWS Just after Tycho moved to Prague in 1599, he had the good fortune to meet the 29-year-old German mathematician named Johann Kepler. Tycho, recognizing his talents, took him on as his assistant. Kepler found Tycho as exasperating as everyone else did, and several times threatened to leave. But before he could carry out his threat, Tycho died, and Kepler was appointed imperial astronomer in his place.

As well as being a fine mathematician, Kepler was deeply religious, and he was convinced there must be an underlying principle governing the motions of the heavens. Inspired by this belief, he set himself the task of finding a set of equations that would fit the movements of the planets, as revealed by Tycho's observations.

Previous astronomers had assumed that the planets' orbits must be circular, and Kepler's first efforts were based on this assumption. Eventually, despairing of making circular orbits work, he tried other figures, and discovered that the observed positions of all the planets, and their moons, could be explained if their orbits were elliptical. This enabled him to formulate his first law:

1. The planets travel in elliptical orbits, with the Sun at one focus.

Further study showed him that the planets travelled fastest in those parts of their orbits which are closest to the Sun, and that this variation in speed was governed by a mathematical relationship he was able to embody in a second law:

2. A planet's radius vector (a line drawn from the planet to the Sun) sweeps out equal areas in equal periods of time.

He next attempted to establish whether there was a relationship between a planet's distance from the Sun and the time it took to complete its orbit, and he discovered that this relationship too could be described in simple formula, which he embodied in his third law:

3. The square of a planet's orbital period is proportional to the cube of its average distance from the Sun.

With this third law, Kepler placed a powerful weapon in the hands of astronomers. Since the Earth's distance from the Sun, and its orbital period, were known with accuracy, one had only to observe the duration of a planet's orbital period in order to calculate that planet's precise distance from the Sun.

Kepler's search for principles underlying the movements of the planets consumed 20 years of his life. The outcome was a triumph, and it changed astronomy utterly. Copernicus had engineered a revolution, and had dethroned the Earth from its place as the centre of the universe. But even Copernicus had been a slave to the idea of the circle as the only conceivable path a heavenly body could follow, and in his universe the planets and the stars were still attached to a series of heavenly spheres. Kepler swept all this away. His planets were freely moving bodies, following their courses through empty space, with no crystal spheres separating them from the stars beyond. And those courses were ellipses, whose shapes were determined by simple mathematical relationships.

Brilliant mathematician and bold speculator though he was, one answer was beyond his reach. As he studied the varying speeds of the planets as they traversed their orbits, and the strict relationship between the period of those orbits and their distance from the Sun, he became convinced that some power emanating from the Sun was the key to their movements. But what that power was, and what linked it to the laws he had uncovered, he was unable to guess. It would take an even greater mathematician, Isaac Newton, who was born 12 years after Kepler died, to finish the task he had begun so brilliantly.

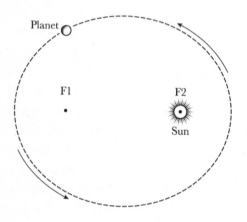

Figure 6. Kepler's First Law
The planets' orbits are ellipses, with the Sun at one focus.

THE CIRCULATION OF THE BLOOD William Harvey was born in Folkestone, Kent, in 1578, the eldest son of a rich merchant. He attended Cambridge University, and went from there to Padua in Italy, where he studied medicine under the anatomist Hieronymus Fabricius. The medical school in Padua led the world in the study of anatomy, and the knowledge Harvey acquired there was to be the foundation of his later achievements.

On his return to England, Harvey set up as a doctor. He enjoyed such success that he became court physician to James I, and later to Charles I. This gave him the leisure to continue his studies of anatomy, and to conduct a systematic programme of research into the workings of the heart and the blood.

Harvey established that the heart was a muscle, which pumped out blood by contraction. Having calculated that the quantity of blood it pumped out in an hour exceeded the weight of the entire body, he concluded that it could not possibly manufacture new blood, and destroy old, at such a rate, and that this must be the same blood re-circulating. He observed the one-way valves between the upper and lower chambers of the heart, and the similar valves in the veins, and he deduced, correctly, that the blood in the veins flowed only towards the heart, not away from it. He tied up arteries, and watched the blood pile up on the side nearest the heart. Every experiment he performed reinforced his belief that – rather than flowing back and forth between the heart and the extremities, or being constantly replenished and dispersed – the blood circulated around the body.

Harvey began lecturing on his theories in 1616, but it was not until 1628, when he was 50 years of age, that he published (in Latin) the book that made him famous: *Of the Motion of the Heart and Blood in Animals*. Nowadays it is accepted as a classic of science history, but on its first publication it met with ridicule. The problem was that, despite years of painstaking dissection, he could not explain how the blood moved from the arteries to the veins. Without a microscope, he was unable to see the very fine vessels we call 'capillaries'.

The Greek philosopher Galen had speculated that the arteries and the veins were self-contained systems carrying different fluids, and this had been the accepted wisdom for 1,400 years. Harvey disputed this, suggesting that the blood from the arteries flowed into the veins, through vessels so fine as to be invisible. But it was only a guess, and his contemporaries needed stronger evidence if they were to reject the teaching of the great Galen. They poured scorn on Harvey's ideas; his scientific reputation and his medical practice suffered as a result.

As the years went by, opinion changed in his favour. His reputation recovered, and he lived to see his ideas achieve general acceptance. But it was not until 1661, four years after his death, that an Italian physiologist, Marcello Malpighi of the university of Bologna, looked into his microscope, discovered the capillaries in the lungs of a frog, and proved that Harvey had been right all along.

It is sometimes claimed that the true discoverers of the circulation of the blood were the group of gifted anatomists who preceded Harvey at Padua, or the thirteenth-century Arab physician Ibn al-Nafic al Quarashi, or even the Chinese scholars who wrote on the subject hundreds of years before Harvey was born. There is no argument that ancient Chinese science, passed on by Arab writers, inspired the anatomists of Renaissance Italy, and that these in turn inspired Harvey's own researches. But Chinese theorizing was a mishmash of valid physiological insights and way-out ideas of mysterious 'essences'. And although it was his Italian predecessors who discovered the valves in the veins, and the 'small circulation' of the blood between the heart and the lungs, it was Harvey who explained the one-way function of the valves, and the all-important 'large circulation' from the heart back to the heart. He deserves the praise that English historians of science have lavished upon him, even if, in honouring him, we should remember just how much he owed to a great Italian university, and the education he received there.

THE HEART AND ITS FUNCTION Although Harvey established the *fact* of the circulation of the blood, he did not discover its *function*. It would be another 30 years before another English physician, Richard Lower, would demonstrate that dark venous blood became red arterial blood on contact with the air. And it would be a century after that before the French chemist Lavoisier would identify the vital element – oxygen – that was the key to the blood's role in the body's metabolism.

The heart is the body's largest muscle – and it needs to be. Its workload and its stamina are phenomenal. It is the most efficient pump ever created. Every minute of every day, throughout an entire adult lifetime, depending on the demands made upon it, it pumps out, and reprocesses anything between 4 and 24 litres/8.5 and 51 pints of blood. It not only circulates this huge quantity; it sends it on its way recharged with the oxygen essential to the functioning of all the body's organs, the brain included. And it does so from a reservoir of blood that amounts to only about 5 litres/10.5 pints in total.

THE ROYAL SOCIETY The period of English history between 1660 and 1700 produced one of the greatest flowerings of scientific discovery ever seen, and some of the credit belongs to a man who was not a scientist at all.

In 1660, the Puritan republic created after the English Civil War came to an end with the restoration of the monarchy under Charles II. Charles was the son of Charles I, the king the republic had beheaded. He had spent the previous 15 years in exile, and was determined not to repeat the experience. His main concern was therefore to avoid the bitter confrontation between religious factions that had embroiled his forebears. Political necessity and his own easy-going temperament combined to create an atmosphere of tolerance and freedom of thought in which scientific enquiry flourished. In 1662, he took pleasure in putting his name to a charter incorporating the Royal Society of London which, over the next half-century, became a fountain of scientific discovery.

Charles' Royal Society was not the first body of its kind. Florence had its Accademia del Cimento (Academy of Testing), itself a successor to the Lyncean Society of Galileo's day. But this folded a few years later, leaving Italy without a comparable body. Paris's Académie des Sciences had been incorporated five years earlier, but it was a public institution, funded by the state. The Royal Society was different. It was a gentleman's club, not a government creation. It had its origins in an informal group of scientifically minded men (no women – perish the thought!), who had been meeting and corresponding with scientists elsewhere since 1645. Despite its royal charter, it was not a public body, and enjoyed no state subsidy. This made it a perfect vehicle for what a later age would call networking. In the

50 years following its incorporation, its membership was a roll-call not only of the nation's own scientific luminaries – Newton, Hooke, Halley, Wren, and the rest – but also of many of Europe's most original scientists: men of the calibre of Leeuwenhoek and Huygens, whose regular correspondence informed its deliberations. In such a forum, and in such an atmosphere, it is not surprising that science acquired a new momentum.

NEWTON'S CONCEPT OF GRAVITY In 1665, an out-break of plague terrorized the city of London, driving those who were able to escape to seek refuge in the country. Many public institutions in London and elsewhere closed down for the duration of the epidemic. One of these was the university of Cambridge; it was thus that a 22-year-old scholar named Isaac Newton found himself back on his mother's farm in Lincolnshire. Newton had enrolled at Cambridge in the year after Charles II had assumed the throne. He had graduated without distinction, but he was a mathematician of extraordinary powers; and the mental activity he engaged in during the next 18 months, in the quiet of his childhood home, formed the foundation of arguably the greatest single piece of work in the whole history of science.

Like some other scientists of genius – Darwin and Einstein among them – Newton gave no sign of intellectual brilliance in his youth. But an uncle, who was a fellow of Trinity College, Cambridge, persuaded his mother that the boy would benefit from a good education. He obtained a place for Newton at Trinity as a subsizar, a student who paid for his keep by acting as a servant to students from more affluent backgrounds.

Although he did not distinguish himself in his examinations, Newton had already done some highly original work in mathematics. Within months of returning home he went on to develop what we know as the binomial theorem, and made considerable progress towards the invention of differential and integral calculus. At the same time he was pursuing research into the subject of optics (*see page 71*). As if this were not enough to occupy his mind, he was simultaneously working on celestial mechanics – the study of the motions of the heavenly bodies.

Newton addressed himself to a standard objection raised by people who rejected the idea of a rotating Earth: if the Earth was spinning, why didn't centrifugal force fling objects on the surface off into space? Newton's answer was that there must be, as evidenced by the behaviour of falling objects, an even stronger force attracting them towards the Earth. By observing the fall of a long pendulum, he was able to calculate the downward acceleration at the surface of the Earth. When he calculated the value of the centrifugal force operating at the Earth's surface, it came to only $\frac{1}{300}$th of the downward acceleration.

He next considered the problem of the Moon, and why it similarly did not fly off into space. He concluded that there must be a similar attractive force operating on the Moon, sufficient to keep it in its orbit. When he reduced his attractive force to its equivalent at the distance of the Moon, in accordance with an inverse square rule, he found that it 'pretty well' (his words) matched his calculation of the centrifugal force generated by the Moon's motion around the Earth. In other words, the downward pull of the Earth was just sufficient to balance the Moon's tendency to fly away into space.

Newton claimed that he was led to his theory by reflecting on the fall of an apple in his mother's orchard. We have no evidence that this was true, and it may just be a story he made up. Worse still – given his pathological inability to credit other scientists with any part in his discoveries – it may have been a ploy to shut them out. But it is a good story, so maybe we should give him the benefit of the doubt.

Whatever its source, the inspiration was remarkable. What Newton had done was discover a principle that was capable of uniting in one law the behaviour of earthly objects like apples, and heavenly objects like the Moon. To do this he had had to reject thinking that dated back to Aristotle: that the Earth and the heavens were separate domains governed by different rules. He also had to employ a concept that seemed magical rather than scientific: the concept of *a force acting at a distance*, an invisible force that was capable of holding the Moon in orbit around the Earth and, by implication, a distant planet in orbit around the Sun. It is difficult now, accustomed as we are to the idea of a force of gravity pervading the whole of space, to appreciate how revolutionary such an idea was.

What Newton did was more than just think up the *idea* of a force of gravitational attraction operating across vast distances; he demonstrated the mathematics underlying it. And he embodied his profound conclusions in statements so simple that any educated person could understand them. The poet Alexander Pope summed up Newton's achievement in a couplet that, though familiar, is difficult to better in terms of its description of the impact Newton's laws had on his contemporaries:

Nature and nature's laws lay hid in night;
God said 'Let Newton be', and all was light.

UNIVERSAL GRAVITATION Newton's theory was constructed on foundations laid down by scientists who preceded him, in particular Kepler and Galileo. Kepler established the nature of the orbits of the planets around the Sun, and the mathematical formulae that described them. Galileo uncovered the mathematics of falling bodies. Newton's achievement in relation to gravity was twofold. He discovered the underlying principle that accounted for both Kepler's laws *and* Galileo's mechanics, and he provided mathematical proof that Kepler's laws were not special cases applying to a particular group of planets, but the inevitable consequence of a force that operated throughout the universe.

His law of universal gravitation can be embodied in a simple statement:

The gravitational attraction between two bodies is
a. proportional to the product of their masses,
and
b. inversely proportional to the square of the distance between them.

This short statement has provided an unparalleled understanding of natural processes. It has enabled us to explain the tides, weigh the stars, and walk on the Moon.

NEWTON'S *PRINCIPIA* Newton had established the basics of his gravitational theory by the time he was 25, but it would be another 20 years before he would publish the book in which he made it known to the world. In 1667, he returned to

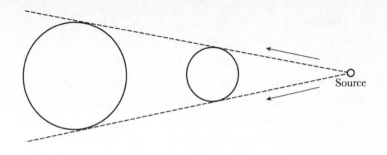

Figure 7. Newton's Inverse Square Law
The gravitational force exerted by any body diffuses, like light, so that at twice the distance it is spread over four times the area, and its strength is reduced to a quarter of what it is at the shorter distance.

Cambridge, and 2 years later he was appointed Lucasian professor of mathematics. The post involved very little teaching; which was fortunate, as he was an uninspiring lecturer.

Newton's theory can be summarized in the 25 words quoted above. But its details, and the book in which it was expounded, defeated all but a handful of the finest intellects of his day. One aristocratic would-be reader offered a reward of £500 (£50,000 in today's money) to anyone who could explain it to him.

Exactly why Newton put his gravitational work on hold for the best part of 20 years is unclear. It is true that he was not very interested in communicating the results of his researches to the outside world. He himself would have said that he pursued science for his own satisfaction, not for public renown. But he was happy to be elected a member of the Royal Society in 1672. And he lost no time in communicating to his fellow members some of his discoveries in the field of optics.

He had other interests competing for his attention, including alchemy, and some weird investigations into the hidden meaning of certain biblical texts. But it is hard to believe that, if he had been confident of the soundness of his gravitational theory during these years, he would have been reluctant to bask in the glory of being its discoverer. Great man though he was, his best friend – if he had had one – would have hesitated to call him a nice person. He had a mean, vindictive streak, and a character that could not tolerate being less than indisputably right. It would seem that at least some of his secretiveness was attributable to insecurity, generated by a discrepancy between his calculations and the actual forces observed.

Whatever the reasons, it was only as a result of pressure from other members of the Royal Society that he addressed the subject afresh, producing the fully worked-out treatment of his gravitational theory embodied in his book, *The Mathematical Principles of Natural Philosophy,* which was published in 1687. As was usual in his day, it was written in Latin; it is still known as 'Newton's *Principia*'. It would be another 5 years before he would adopt the English word 'gravity' to describe his universal force. And it was not until 1729, 42 years after its publication, and 2 years after his death, that his book made its appearance in his native tongue.

NEWTON AND THE CAT FLAP Einstein considered Newton to be the greatest scientist who has ever lived, so the rest of us had better believe it. It's comforting to know that even his colossal intellect did not always lead him to the most economical solution. He once cut a hole in his door, so that his cat could come and go without disturbing his concentration. When the cat had kittens, he added smaller holes for them.

MASS AND WEIGHT   One thing that Newton's work made clear is the difference between *mass* and *weight*. Mass is the total amount of matter in a body. Weight is a measure of the effect of gravity on that material. Astronauts with a mass of 70 kilograms/150 pounds on the surface of the Earth will have the same mass on the surface of the Moon. But whereas their weight on Earth will be 70 kilograms/150 pounds, their weight on the Moon will be only 14 kilograms/30 pounds. And while they are in orbit, they may weigh nothing at all.

But they need to be careful how they weigh themselves. If they use the bathroom scales, there will be no problem. Bathroom scales measure weight. But if they make the mistake of weighing themselves in an outsize balance, with themselves in one pan and a 70-kilogram/150-pound weight in the other, the result will be the same on the Moon as it was back home. Balances measure mass.

EDMUND HALLEY'S COMET   Given his limited interest in communicating his discoveries to his fellow scientists, it is safe to say that, left to him, Newton's *Principia* would not have been written. That it *was* written, and published, is down to one of the few men who got close to this prickly and paranoid genius.

Edmund Halley was a young man of private means who attended Oxford University: arriving with a telescope 7-metres/24-feet long, but leaving without a degree. While an undergraduate, he entered into correspondence with John Flamsteed, the Astronomer Royal, about errors in some published astronomical tables. Flamsteed was engaged in producing an improved catalogue of the stars in the northern sky, and Halley conceived the idea of producing a similar catalogue of the southern sky. He accordingly abandoned his studies, and sailed to St Helena, largely at his father's expense, as the leader of an astronomical survey. On his return to London, still only 21, he was elected to membership of the Royal Society.

Halley was a scientist of the first rank, with many achievements to his name. He compiled the world's first sea charts to show magnetic variation. He was the first person to establish that some stars had their own *proper motion*. He did this by comparing the position of three stars – Sirius, Procyon, and Arcturus – in his own day, with the positions recorded in the past by Ptolemy and Tycho Brahe. He concluded that the differences he uncovered were the result of the stars in question having moved – a suggestion that would not receive experimental proof until nearly two centuries later.

Halley's great service to science was persuading Newton to write his *Principia*. But his immortality was assured by something that happened after he died. When he was 25, he witnessed the great comet of 1682. This led him to study notable comets of the past, and he was struck by the similarity of the path of the 1682 comet to comets recorded in 1456, 1531 and 1607. In 1705 he published a paper suggesting that these were successive visits of a comet that orbited the Sun in an elongated orbit with a period of about 75 years. This may seem obvious now, but in 1705, only 18 years after the publication of the *Principia,* it seemed to many people to be more than a little far-fetched. Halley predicted that the comet would return in 1758.

He lived to be 85 but not long enough to see his forecast put to the test. In 1758, 16 years after his death, his comet arrived. It was a sensational event: the first demonstration of the predictive power of Newton's laws, and the coming-of-age of the science of celestial mechanics.

Halley's Comet, as it is known, has never failed since. It last appeared in 1986, and astronomers look forward confidently to its next visit in 2061. It will not be impressive, because the comet will not be particularly close to the Earth on that occasion. But anyone around in 2137 is promised a brilliant spectacle.

The comet is presently about halfway through its outward journey. In 2003, when it was about 4 billion kilometres/2.5 billion miles distant, it was picked up by the Very Large Telescope (VLT) at Paranal, in Chile. With its nucleus 10 kilometres/6 miles across, and its albedo (sunlight-reflecting power) of only 4 per cent, it was then about a billion times fainter than the faintest naked-eye star. Such is the power of modern telescopes.

THE COLOURS OF LIGHT   During those 18 months on his mother's farm, Newton conducted a number of investigations into the behaviour of light. These led him to suspect that sunlight was composed of different kinds of light. Knowing that a glass prism (a length of glass with a triangular cross-section) was capable of bending light rays, he decided to see whether such a prism could bring about the separation he was looking for. He was not the first person to investigate the effect of a prism on sunlight. But other experimenters had only been interested in the bending effect, and had focused the bent rays on a flat surface close to the prism, with the result that they had observed only a small pool of white light. Newton had the inspiration of focusing the light thrown by *his* prism on a wall 7 metres/22 feet away, and he was rewarded with a band of light that displayed all the colours of the rainbow.

Newton realized that this did not prove that the colours had been contained within the light before it entered the prism. They could have been created within the prism itself. He therefore placed a lens in the path of the light emerging from the prism, and focused the rays onto a single point. When he did so, he discovered that this reconstituted the colours into white light again. This led him to three conclusions:

1. that sunlight was a mixture of light of seven different colours,

2. that the *refractive index* – the bending effect – of glass was greater for some colours than for others, *and*

3. that the increase in refraction was progressive across the spectrum.

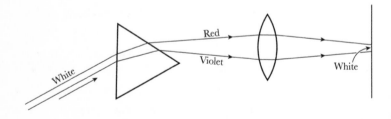

Figure 8. Newton's Experiment with Sunlight
When sunlight strikes a prism, the differing angle of refraction of light of various wavelengths splits it into its constituent colours. When these are refocused by a lens, the original white light is reconstituted.

Newton was not the first person to notice the ability of a glass object to render white light into a range of colours. It was a phenomenon that had been familiar almost since the invention of clear glass. But previous observers had assumed the light was transformed by its passage through the glass. Newton's achievement was to prove that the colours were inherent in the light itself. It was a discovery that yielded no application in his own day; but it provided the basis of a scientific technique – spectroscopy – that would later revolutionize the practice of astronomy.

THE RAINBOW   The significance of Newton's prism experiment was that it proved for the first time that the colours of the rainbow were contained within the Sun's light, not a transformation of it. The rainbow itself had been explained before Newton was born.

Rainbows are caused by the refraction of sunlight as it passes through water droplets, in a shower of rain, or in a waterfall. Each droplet acts like a prism. Light from the Sun behind the spectator is reflected from the far wall of the droplet, and emerges as a band of colours which have been separated on their passage through the raindrop's walls. The individual colours that make up the rainbow come from droplets whose different positions in the sky cause them to deliver a different part of the spectrum to the spectator's eye. The arc is part of a circle of refracted light, with a centre directly opposite the Sun's position in the sky. How much of the circle is visible depends on how low the Sun is, and how high the observer is. Sometimes a second rainbow is visible, with its colours reversed. This second rainbow is caused by light striking other droplets at a steeper angle, and taking a different path within them.

THE ELECTROMAGNETIC SPECTRUM We know now that the Newton's spectrum is just one band – the visible part – of a much wider spectrum of electromagnetic radiation, most of which is invisible to human eyes:

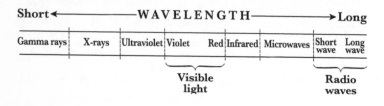

Figure 9. The Spectrum of Electromagnetic Radiation

HOW WE SEE COLOUR Colour is so much a part of everyday experience – for those of us who are not colour-blind – that it is hard to believe that it only exists inside our heads. When we talk about something being blue, or green, or yellow, we are not describing a fact of Nature, we are merely giving a name to a sensation we experience, just as we do when we describe something as being 'rather hot', or 'very cold'. The spectrum of visible light is a continuous scale of electromagnetic radiation of varying wavelengths. The retina of the human eye contains about six million light-sensitive cells called 'cones', which respond to specific wavelengths of light. About two-thirds of these are 'red' cones, which register light in a narrow band in what we call the red area of the spectrum. Just under a third of them are 'green' cones, and the remainder – a mere 2 per cent

– are 'blue cones'. The brain defines the colour of any object by comparing the intensity of the messages received from the three types of cell. It is for this reason that we perceive these three as the 'primary' colours, and call other sensations, which are a sort of average score, as 'secondary' colours.

Because colour is a mental sensation, not a physical fact, the eye can be tricked into 'seeing' what one might be tempted to call natural colours, which are actually created of completely different wavelengths. The most striking example of this kind of deception is colour television. A television picture is made up of the colours magenta, yellow, and cyan. The wavelengths of these colours are quite different from the primary colours of red, green, and blue, but the brain's 'averaging' mechanism nonetheless registers a colour image closely corresponding to what it would record if the eye were where the camera lens is.

DIFFERENT KINDS OF LIGHT If you go out on a sunny day you can't be in any doubt that the Sun's light comes from a hot place. The source of the Sun's light, and of most of the light in the universe, is *incandescence,* the light given out by hot substances. This was clear to Newton, as it had been to Aristotle. When Newton's contemporaries, seated in the noonday sun, reflected on the significance of his inverse square rule, and the enormous distance separating the Earth from the Sun, they realized that the Sun must be very hot indeed. But as to what might be capable of giving rise to heat on such a scale, over such enormous periods of time, they were no wiser than the Ancient Egyptians. It would be another two centuries before it would become known that the Sun's fires were fuelled by a process unknown on Earth.

Incandescence is the most important source of light on Earth, but it is not the only one. There is also the light emitted by cold substances. This is called *luminescence*. It comes in four varieties:

1. *Bioluminescence* – the light emitted by creatures such as fireflies. The source is a chemical reaction within the animals' bodies.

2. *Phosphorescence* – a gradual release of previously stored energy by substances such as phosphorescent paints. The energy is absorbed in sunlight, and then gradually released. The release is a continuous process, but becomes noticeable in the dark.

3. *Fluorescence* – the rapid release, in the form of visible light, of energy that has been absorbed from ultraviolet light. Because fluorescent materials obtain their light energy from ultraviolet light, they can glow in the dark.

4. *Triboluminescence* – a form of light briefly released when some kinds of crystals are crushed.

THE SPEED OF LIGHT A question Galileo had asked himself, but failed to answer was, 'Does light have a finite speed?' Aristotle had said no; and if Aristotle said no, that was good enough for most seventeenth-century astronomers. In an attempt to find out if Aristotle was right, Galileo experimented with light signals between two hills. He stationed himself on one hill, and on another he placed an assistant, who was instructed to signal back when he observed a signal from Galileo. The idea was that the delay in receiving the acknowledgement of his own signal would indicate the time taken by light to traverse the double distance. The principle was sound, but the delay was extremely slight, and could easily have been attributable to delayed reaction on the part of Galileo or his assistant. In 1672, a Danish astronomer named Ole (Olaus) Roemer, equipped only with a telescope, succeeded where Galileo had failed.

Roemer was the son of a ship-owner. He was born in Aarhus, the second city of Denmark, in 1644. He studied astronomy in Copenhagen; in 1671, when he was 27 years of age, he was invited to Paris to work as an assistant to the French astronomer, Jean Picard. He remained there until 1681, when he returned to his native Denmark as astronomer royal.

Picard, who was professor of astronomy at the Collège de France, would later go on to make the first precise measurement of the circumference of the Earth, using the same method as Eratosthenes, but based on the inclination of a star, not the Sun. The use of a star, which is a pin-point of light, rather than the Sun, yielded much greater accuracy; it was Picard's measurement that would in due course enable Newton to confirm the validity of the calculations underlying his theory of universal gravitation.

Picard had supervised the collection, by another of his assistants, the Italian Giovanni Cassini, of a series of observations of the motions of the satellites of Jupiter. These made it theoretically possible to forecast exactly when a satellite would be eclipsed as it passed behind the planet. Roemer noticed a pattern in these observations. When the Earth and Jupiter were on the same side of the Sun – that is, when they were close together – the eclipses tended to be early. When the Earth and Jupiter were on opposite sides of the Sun, the eclipses were always late.

In a moment of inspiration, Roemer guessed that the discrepancies were due to the time taken by light to travel across the intervening space. When the Earth and Jupiter are on the same side of the Sun, they are only 650 million kilometres/400 million miles apart. When they are on opposite sides of the Sun, the separation is nearly 950 million kilometres/600 million miles. Roemer concluded that the difference in time between the most premature and the longest delayed observations was the time it took sunlight reflected from the satellites' surfaces to cover the extra 300 million kilometres/200 million miles. Applying this reasoning, he calculated the speed of light to be 220,000 kilometres/140,000 miles per second. He announced his result at a meeting of the Academy of Sciences in Paris in 1676. It made no great impact on his contemporaries, even though some eminent men of science – including Picard, Huygens and Newton – were inclined to support his calculations. More recent measurements have established that the correct figure is 300,000 kilometres/186,300 miles per second, making Roemer's estimate about 25 per cent understated. This is no reflection on either his logic or his arithmetic. His calculations were based on contemporary estimates of the size of the Earth's orbit. Had he been able to use the correct figure, his sums would have given him the correct answer.

Anyone who is tempted to think that technology is always applied science, and never the other way round, should consider that this proof of the finite nature of the speed of light – a discovery vital to the progress of physics and cosmology in the twentieth century – would not have been possible without the invention of the telescope.

LOOKING BACK IN TIME   The proof of the finite velocity of light presented humanity with a stunning new fact: that when we look out into space, we are looking back in time. When we look at the Moon, we are seeing it as it was 1.5 seconds ago. When we look at the Sun, we are seeing it as it was 8 minutes ago. When we look at the nearest star, we are seeing it as it was 4 years ago. When we look at the Great Nebula in Andromeda – the nearest galaxy to our own – we are seeing it as it was 2 million years ago. And when we look at the most distant visible galaxy, we are seeing it as it was when the universe was young, 10 billion years ago.

ESCAPE VELOCITY   When the crew of Apollo 11 landed on the Moon in 1969, one of their first acts was to unfurl the Stars and Stripes. It would have been fitting if their last act had been to write in the dust the words 'Isaac Newton Was Here'. Space travel really *is* applied science; and its technology is rooted in two concepts central to Newton's planetary theory: *escape velocity* and *momentum*.

If an arrow is fired upwards, the height it attains depends upon the speed with which it is projected. From the moment it leaves the bow, the Earth's gravity is slowing it down; it ceases to climb when the pull of gravity exceeds its own upward momentum. The same is true of a spaceship, which is an arrow fired at the stars. But while an arrow *must* fall back to Earth, a spaceship has a choice. This is not because it has an unlimited supply of fuel. It is because there is a velocity that, once attained, enables any object to escape forever from the pull of the planet from which it is projected. The value of this *escape velocity* is different for every planet.

The strength of gravity at the surface of a planet depends upon two things: the planet's mass, and the distance from its centre. As an object moves away from a planet's surface, the pull of gravity gets less, in accordance with Newton's inverse square law. An object at the surface of the Earth is 6,400 kilometres/4,000 miles from the Earth's centre. At 6,400 kilometres/4,000 miles above sea level, the strength of gravity is reduced by 75 per cent. (Doubling the distance from the Earth's centre reduces the strength of gravity to only a quarter of what it was at the surface.) At a height of 32,000 kilometres/20,000 miles, it has only $\frac{1}{25}$th (4 per cent) of its strength at sea level. Because the strength of gravity lessens as one moves away from the Earth, there is a launch speed that is fast enough to ensure that gravity will never succeed in stopping a spaceship. At the Earth's surface, this *escape velocity* is 11 kilometres/7 miles per second.

The concept of escape velocity is widely misunderstood. It is not, in practice, necessary for a craft to be launched at anything like this speed. In fact, as a moment's reflection will make clear, a spaceship needs, in theory, to receive only a gentle, sustained lift sufficient to keep it slowly rising, and it is bound to escape. The trouble is that its journey would be very expensively prolonged, even if its fuel did not run out. What happens in practice is that the ship is subjected to a gradually increasing acceleration for 5 minutes or so, after which it will be, say, 1,600 kilometres/1,000 miles above the Earth. At this point, having reached the escape velocity of 6 kilometres/4 miles per second that applies at that height, it is injected into a circular orbit, from which it embarks on its planned trajectory.

Another concept underlying the technology of space travel is *momentum*. This is not the same as velocity. Momentum is mass multiplied by velocity. If two identical trucks are running downhill at 80 kilometres/50 miles an hour, the truck with the heavier load will have the greater momentum; it will require a more violent application of the brakes to slow it down. If the brakes aren't applied, the truck with the greater momentum will demonstrate the extra energy its motion embodies by the damage it does to itself and whatever it crashes into. In the same way, it requires more energy to push a heavily laden spaceship up to its escape velocity than it does to impart the same speed to an unladen one. Early space engineers spent a lot of time working out how to escape from a vicious circle in which more acceleration called for more fuel, which necessitated a bigger ship, which called for more energy, which called for more fuel, and so on. It was this quandary that dictated a design that made possible the successive jettisoning of large sections of the vehicle as its fuel was used up.

The good news for the planners of the first lunar mission was that the Moon's escape velocity is much less than the Earth's. Although the surface of the Moon is closer to its centre than is the case with the Earth, the Moon's mass is only $\frac{1}{80}$th of the Earth's. As a result, the strength of gravity is much less on the Moon. Just as the lunar explorers were able to jump higher than they could at home, despite their heavy gear, so their lunar module needed to attain a much lower take-off speed in order to leave the Moon, than the mother ship had needed to escape from the Earth. The opposite is true of the larger planets. The strength of gravity at the surface of Jupiter is $2\frac{1}{2}$ times that on the Earth, and the escape velocity is 60 kilometres/37 miles a second.

LEAVING THE MOON *Question:* If the escape velocity of the Earth is 11 kilometres/7 miles per second, the mass of the Moon is $\frac{1}{80}$th of that of the Earth, and the radius of the Moon is $\frac{1}{4}$ of the radius of the Earth, what is the escape velocity of the Moon?

Answer: Other things being equal, the mass of the Moon would cause its power of gravitational attraction to be $\frac{1}{80}$th of that of the Earth. But the fact that its radius is only $\frac{1}{4}$ of that of the Earth would, by itself, cause the force of gravity at its surface to be 16 times greater than at the surface of the Earth. The force of gravity on the Moon must therefore be $\frac{1}{80}$th multiplied by 16, which is $\frac{1}{5}$th, of that on the Earth. This means that the Moon's escape velocity must be $\frac{1}{5}$th of 11 kilometres/7 miles per second, which is 2.4 kilometres/1.5 miles per second.

CALCULATING ORBITS One of the first lessons a hunter has to learn is that it is no use aiming his gun at a running deer. By the time the bullet has reached the place where the deer was, the deer will be somewhere else. Similarly, it is no use aiming a spaceship directly at the passing Moon. By the time it gets there, the Moon won't be. The science of space travel is in this respect, the science of hunting: you have to aim for the spot where you hope the moving target will be.

But space travel is more difficult than taking pot shots at running deer from the front porch. In this case, the porch is moving too. The spaceship is already moving fast before count-down begins. Like everything on the Earth's surface, it is travelling in a circle at over 1,600 kilometres/1,000 miles an hour. It is also moving – at 80,000 kilometres/50,000 miles an hour – in the direction of the Earth's orbit. This is no problem for the astronaut boarding the spacecraft, because he is moving at the same speed, along the same path. But it is a problem for the flight planner. Unlike the seated hunter, he cannot aim his projectile straight at the spot where he expects the Moon to be. He has to incorporate the motion the spaceship shares with the Earth's surface into a flight path capable of achieving a rendezvous at the point where the Moon will be when that path has been traversed. The solution to his problem is an outward spiral, combining the Earth's motion and the ship's upward movement, which the ship will follow until it

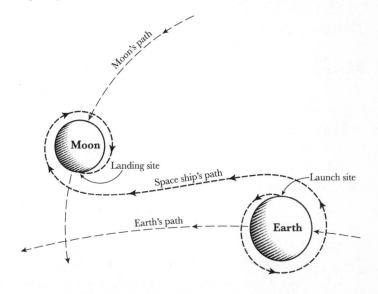

Figure 10. Simplified Illustration of a Journey to the Moon
N.B. This shows the principle involved, not an actual journey.

falls into the welcoming arms of the Moon's gravity, and travels in an inward spiral to its landing site. It is the sort of calculation for which a computer is quite handy.

TYPES OF MOMENTUM The momentum possessed by a truck running downhill is *linear momentum*. The momentum possessed by systems such as the Earth or Moon is *angular momentum*. All motion conforms to a rule called the *conservation of momentum*, which states that momentum can be transferred, but it cannot be destroyed. If one billiard ball strikes another, the total momentum possessed by the two balls after the collision must equal the momentum previously possessed by the first moving ball. It is the same principle that dictates that the Moon's orbit must enlarge as the Earth slows down, so that the angular momentum of the system is conserved.

HUYGENS AND THE PENDULUM CLOCK Between the publication of Copernicus' book in 1543 and the death of Galileo in 1642, physics and astronomy made enormous advances. But, by the middle of the seventeenth century, both sciences were confronted by a problem that threatened to bring further advance to a halt. The problem was one of measurement: the measurement of time. Galileo had done remarkable work using water clocks and the beating of his pulse. But water clocks were anything but precision instruments; and the human heartbeat was useless for measuring elapsed time. Mechanical clocks of a sort had been in use since the fourteenth century; but they basically measured hours, not minutes – and certainly not seconds. And even within their capabilities, they often became inaccurate to the tune of half an hour in the course of a day. The man who found the solution to the problem, and gave science one of its most essential tools, was a Dutch physicist and astronomer named Christiaan Huygens.

Huygens, who was 14 years older than Newton, was born in The Hague in 1629. He was the son of a senior official in the Dutch government, and he was originally expected to become a diplomat. He spent 2 years studying mathematics and law at the university of Leiden, followed by a further 2 years studying law at the university of Breda. But his first love had always been mathematics; and when he was 20 years old, he abandoned his diplomatic ambitions and, with his father's blessing and financial support, he decided to devote himself to science. For the next 17 years he worked away quietly at his researches in the peace and security of the family home. When he was 26, he discovered a way of grinding lenses that produced images of far superior quality to any previously obtained. He incorporated his lenses into a telescope of his own design, with which he made a number of discoveries, including the Great Nebula in Orion, the rings encircling the planet Saturn, and a satellite of the same planet, which he named Titan.

Huygens' greatest achievement was his invention of the pendulum clock. He was not the first to think of using the swing of a pendulum to regulate the motion of a clock. Galileo had noted the consistency with which a pendulum swung; before he died he had made drawings of a possible clock mechanism controlled by a pendulum. Other scientists had endeavoured to follow up Galileo's discovery without success. It was Huygens who made the crucial breakthrough, when he recognized that for a pendulum to perform with complete accuracy, the arc through which it swung needed to be not the arc of a circle, but a curve of a slightly different shape: a 'cycloidal arc'. By fitting attachments of his own design to the fulcrum of a pendulum, he was able to make it swing in the required arc, and by employing precisely cut wheels and escapement, he was able to transfer the accuracy and consistency of his weight-driven pendulums to the mechanism of his clocks. He published details of the mechanism in his book *The Clock* in 1658; but his great work was the treatise entitled *The Oscillations of Pendulums,* which was published in 1673. This book, which dealt in detail with the mathematics of pendulums, concluded with a set of theorems concerning centrifugal force in circular motion that helped Newton in the formulation of his theory of universal gravitation.

HUYGENS AND NEWTON   Huygens was perhaps the greatest of the many fine scientists whose work was overshadowed by Newton's. He was particularly unfortunate in the matter of the behaviour of light, where his failure to get a hearing for his ideas demonstrated the futility of being right at the wrong time.

Huygens' theory of the nature of light was founded upon discoveries he had made while working on the construction of telescopes. In his *Treatise on Light,* which was composed in 1678 but published in 1690, he had expressed the opinion that light was best understood as a series of waves; and this wave theory enabled him to provide proof of the fundamental laws of optics. It also enabled him to explain the refraction of light, and to predict that light would be found to travel more slowly through dense materials, a prediction that would be confirmed a century later. But by the time the Newton's *Optics* was published in 1704, Newton's reputation was so huge that any contrary opinion to his had no chance of a proper hearing; and the particle theory of light formed part of the conventional wisdom for a century thereafter. It would not be until the early years of the nineteenth century that laboratory experiments would reveal the limitations of Newton's particle theory, and scientists would realize the power of Huygens' alternative interpretation to explain phenomena that Newton's theory could not handle.

We know now that light *is* sometimes best understood as a stream of particles, as Newton proposed. But there are probably more contexts in which it is better understood in the way that Huygens would have had us understand it: as a series of waves spreading out from a source. Anyone who wishes truly to understand light must be ready to switch from one view to the other, as the context requires.

LEEUWENHOEK AND HIS LENSES The title of the first microscopist rightly attaches to the name of the Italian physiologist Marcello Malpighi, whose discovery of capillaries in 1661 proved the validity of Harvey's theory of the circulation of the blood (*see page 61*). But it was a Dutchman, Anton van Leeuwenhoek, who showed the seventeenth century what the microscope was capable of.

Leeuwenhoek was born in 1632, just three years after Huygens, in the town of Delft, where he spent his whole life. He was the owner of a draper's shop, and it was his business that introduced him to lenses: drapers used magnifying glasses to examine cloth. He was also the janitor of the Delft City Hall, and these two sources of income allowed him to indulge his hobby of lens-grinding, a hobby that became a lifelong obsession. By the time he died, he had ground 419 lenses.

Leeuwenhoek was an observer, not a theorist, but as an observer he was unrivalled. He opened the eyes of his contemporaries to the enormous diversity of life. He was the first person to describe the various kinds of plankton in water. He discovered the group of one-celled creatures we call Infusoria. He even discovered and described bacteria: life forms so small that it would be another hundred years before anyone would be able to add anything to what he had to say about them.

He made his first microscope in the 1660s. His instruments were *simple* microscopes, with only one lens. They were basically powerful magnifying glasses, consisting of a tiny, highly convex lens, in the centre of a metal plate. They were held in one hand, and imposed a considerable strain on the eyes. But his lenses were so clear, and so skilfully ground, that their resolving power far exceeded that of the compound microscopes employed by his contemporaries; and they produced images free of the distorting colours (chromatic aberration) that marred theirs'. Of the work of these contemporaries, though, he remained largely ignorant, as he had no Latin, and he could read no language other than his native Dutch.

Despite his humble origin, Leeuwenhoek became a corresponding member of the Royal Society of London, and it was through this one-sided correspondence (he wrote nearly 400 letters, all in Dutch) that his work became known to the outside world. In the last of these letters, he bequeathed the Royal Society 26 of his finest instruments, so that its members could explore this new world for themselves.

In 1677, Leeuwenhoek became the first person to describe spermatazoa. It was one of the most important events in the history of biology, but it did little, in the short term, to advance the understanding of conception. It had been understood for thousands of years that humans, like all sexual creatures, were conceived as a consequence of sexual intercourse. But the mechanics of conception remained as great a mystery in the seventeenth century AD as it had been in the seventeenth century BC. Although Leeuwenhoek was able to describe spermatazoa in detail, he and his contemporaries continued to believe that the function of the female was merely to house and nurture the tiny seed provided by the male. It would not be until 1827 that the German-Russian embryologist von Baer would discover the mammalian egg in the mammalian ovary, and begin the unravelling of the mystery of how human beings are conceived.

Leeuwenhoek became so famous that kings and queens interrupted their journeys to stop off at his shop and peer through his lenses. One lens still survives; this magnified 270 times, enabling him to observe details measuring only $\frac{1}{1,000}$th of a millimetre/$\frac{1}{25,000}$th of an inch in diameter. With these wonderful instruments he established the existence of a world of microscopic life previously undreamed of; and he laid the foundations of several important branches of what was to become the science of biology. He didn't begin his investigations until he was 40. But he lived for another 50 years, and he studied his tiny creatures to the end.

NEWTON AND HIS ERA In the history of science, the seventeenth century is Newton's century. He is like the Sun that blots out every other star in the heavens. But among his contemporaries there were many other great scientists who, without his competition, would have shone very brightly; and there were areas of science – biology, chemistry, and the earth sciences – to which he made no contribution.

It does not detract from his achievement to say that he could not have produced a comparable body of work if he had been working in these areas; or in physics and astronomy 100 years earlier – or later. It truly was the case that he was able to see so far because he was, in his own words, 'standing on the shoulders of giants'. He was able to achieve the synthesis he did because, in astronomy and in physics, there was so much prior discovery for him to synthesize, and it had accumulated to the level where a synthesis was possible.

There were other sciences in the mid-seventeenth century that had barely got off the ground. Biology had been invented by Aristotle, but the intervening years had added little to humanity's understanding of the processes of life. Chemistry and the earth sciences (geology, oceanology, and meteorology) were still closed books.

ALCHEMY AND CHEMISTRY If the young Newton made no progress in the field of chemistry, it was not for want of trying. It is just that, like most investigators of his time, he was chasing a will o' the wisp. It was not chemistry at all – as we understand the word – that he was studying. It was alchemy: an ancient pursuit, followed in the hope of achieving a mastery over the Earth's elements, by transmuting 'base' elements into 'noble' ones. Its dream was the 'philosopher's stone', capable of turning common metals into gold. When he was a student, Newton built a laboratory in his college rooms, where he conducted experiments in transmutation. He was later one of the gullible speculators who expressed a keen interest in the 'secret recipe' of a London company that was incorporated with the avowed aim of multiplying gold.

The trouble with alchemy was that it was guided by no understanding of the structure of matter, or the way in which various substances were related to one another. During the hundreds of years during which it had been practised, it had undoubtedly added something to men's understanding of the materials found in nature, but the amount of effort that had been expended on it was out of all proportion to the small body of knowledge it had produced. Alchemy was a dead end. It took a contemporary of Newton's – another member of that gentleman's club called the Royal Society – to rescue it from its cul-de-sac and to point the way to modern chemistry.

BOYLE'S EXPERIMENTS Robert Boyle was an aristocrat who was born in Waterford, Ireland, in 1627: the fourteenth child, and seventh son, of the wealthy first (and English) Earl of Cork. In 1641, 14-year-old Robert was in Florence with his tutor when he heard of the death of Galileo. This led him to study Galileo's works; and the result was an abiding interest in science. When he returned to England in 1644, he settled in Dorset, but spent much time in his sister's house in London, where he became acquainted with the group of scientists who would later formed the nucleus of the Royal Society. In 1654 he moved to Oxford, and it was there, during the next 14 years, that he conducted many of the experiments that made his reputation.

British historians have been known to refer to Boyle as the 'Father of Chemistry', but that is taking national pride too far. (Given the teamwork involved in scientific discovery, it is doubtful whether anyone should be called the 'Father' of anything. But if the title of 'Father of Chemistry' belongs to anyone, it belongs to the Frenchman Lavoisier, who lived a century later.) Boyle did not create modern chemistry. What he did instead was to free chemistry from some of the dead weight of the past, and clear the way for those who came after, by laying down the crucial principle that chemical facts should be established by experiment, not armchair speculation.

Boyle's experiments, which were carried out with the aid of paid assistants, were many and varied. Using the newly invented air pump, he became the first person to prove Galileo's assertion that, in a vacuum, a feather and a lump of lead would fall at the same speed. He also established that sound did not travel through a vacuum. His most important discovery with the air pump was the principle (still known as *Boyle's law* in English-speaking countries) that the volume occupied by a gas is inversely proportional to the pressure to which it is subjected. That is to say, that if the pressure is doubled, the volume is halved, and so on; also, that if the pressure is removed, the air 'springs' (his word) back to its original volume. Having established that air was compressible, Boyle became convinced that it was composed of small particles separated by empty space. All of these ideas were published in a book with a long-winded title, which is usually referred to as *The Spring of the Air.* It was a book that played a significant part in establishing the idea of the atomic nature of matter.

Boyle's most important book, *The Sceptical Chymist,* was published in 1661, and in the following year he became a founding member of the Royal Society. It was in this book that he put forward the idea that all substances could be divided into acids, alkalis, or neutrals by the use of what we call *indicators.*

CHEMICAL ELEMENTS Perhaps the most significant contribution that Boyle made to the development of what would become the science of chemistry was his concept of a chemical *element.* The word itself was not new. The Greeks, following the philosopher Empedocles, used it to describe what they considered to be the four fundamental substances in the universe: earth, air, fire, and water. These were not in any modern sense scientific concepts. They were more in the nature of mystical essences embodied in all living and non-living matter; and they formed the basis of men's thinking about natural processes for 2,000 years.

Boyle's idea of a chemical element was different. For him, an element was a substance that could not be broken down into other substances. One element might combine with another to form a *compound;* and a compound might be separated into its constituent elements. But the test of what was or was not an element should be decided by practical experiment, not just by thinking about it. This was quite a modern view; and it helped to create the mental universe that later chemists would inhabit. But Boyle himself was unable to shake off the influence of centuries of alchemy. He continued to believe in the possibility of transforming common metals into gold. And he didn't reject the ancient elements – he just wanted them to be subjected to experimental investigation.

DISCOVERING ELEMENTS As scientists began to adopt this new way of thinking, the ancient 'elements' were gradually abandoned, and the term began to be employed in the way we use it today. But the list of substances to which it could be applied was a short one. Only 14 elements, in the modern sense of the word, were recognized at the end of the seventeenth century. Nine of them were metals that had been known since antiquity: gold, silver, copper, lead, zinc, tin, iron, mercury and antimony. Two were non-metallic elements that had also been known to the ancients: carbon and sulphur. Two more were metals that had been discovered in the sixteenth century: bismuth (in Europe) and platinum (in South America). To these 13 had been added one further non-metallic element – phosphorus – which was discovered in urine by Boyle himself in 1680.

Although it is true these 14 elements we recognize today had been identified by the end of the seventeenth century, it is *not* true that they were recognized *as* elements, in the modern sense. When modern chemists talk about elements, they use the word in the sense of the basic ingredients out of which the materials of the world are made. For them, air is a *mixture* of two elements – oxygen and nitrogen – with small quantities of a number of other gases. One of these, carbon dioxide, is for them a *compound* of two elements, carbon and oxygen. This view of chemistry as a collection of recipes, using a small number of basic ingredients, was utterly foreign to the natural philosophers of the seventeenth century. Although they recognized copper, gold and sulphur as 'elements', air was also, for them, an element; and they were still unsure as to whether or not fire was an element too. Unlike the astronomers, who frolicked on the sunlit uplands of Newtonian mechanics, the chemists of 1700 were still fumbling in the dark, seeking for a light to illuminate their path. It would be another hundred years before they would find *their* Newton, and chemistry would be able to take its rightful place among the natural sciences.

INVISIBLE CHEMISTRY The reason chemistry was so late in undergoing its revolution lay in its subject matter. The evidence astronomers needed was in front of their eyes. Even *without* the aid of the telescope, Tycho Brahe had been able to amass the records of planetary movements that enabled Kepler to deduce his laws of planetary motion. *With* its aid, Galileo was able to observe the motions of the satellites of Jupiter. When he used his inclined plane to explore the laws describing falling bodies, he was again able to use the evidence of his eyes to arrive at an underlying truth. When Newton combined Galileo's and Kepler's laws to formulate his law of universal gravitation, he applied the power of *his* mind to the evidence of *their* eyes.

Chemists had no such luck. Chemistry's facts were invisible. Even the microscope, which opened up new worlds to the biologist, was of little use to the chemist. It revealed hitherto unknown structures, and strange new forms of life, but it offered no clue to the nature of the materials of which they were made. It was perfectly reasonable, if one went by appearances, to treat water as being just as much an element as gold, or sulphur. There was no way of knowing, by observing water, that it was a combination of two gases; or, by watching air, that it was a mixture of one of these gases and a third; never mind that the gas which was common to both air and water held the secret of fire. It would require a century of experiments of the kind advocated by Boyle to create the understanding that would turn chemistry from the pursuit of magic to a respectable science. And much of that time was devoted to a wild-goose chase: the search for an 'element' – phlogiston – that did not exist.

THE DANGER OF FAULTY HYPOTHESES People who have not studied the process of scientific discovery often suppose that scientific method consists of collecting facts, and then forming a hypothesis to explain the facts. If this can explain other facts, and support predictions, it is promoted to the status of a theory, which may subsequently be embodied in the form of a physical law. This will be accepted as valid until it is disproved, or modified, as a result of later discoveries. In fact, when great scientists go looking for facts, or when they conduct experiments, they are usually looking for evidence to support, or disprove, a hypothesis that is already at least half-formed in their heads. If they didn't have the makings of a hypothesis, they wouldn't know where to look, or what to look for.

A good example of the true process of scientific discovery is the way in which Charles Darwin arrived at his theory of evolution by natural selection. He didn't spend 20 years assembling facts about the natural world, and then look for a hypothesis to explain them. The similarities between birds on various islands of the Galapagos, and the similarities between past and present life forms in South America, suggested to him that they had arisen as a result of a process of evolution. He *then* spent 20 years collecting a mass of evidence to support his initial hypothesis.

This approach has been the source of some of the most important discoveries in the history of science. Unfortunately, if a hypothesis is both faulty and widely supported, it can result in a lot of misguided effort, which may hold back, rather than advance, science. This is what happened to chemistry in the eighteenth century.

PHLOGISTON – THE ELEMENT THAT NEVER WAS A central concern of chemistry in the eighteenth century was the process of combustion. When substances were heated to the point of incandescence, scientists saw that something – fumes or smoke – was given off; and this was interpreted as the *loss* of part of the original substance. This 'something' that was supposedly given up in the process of combustion was given the name *phlogiston*, a word coined in 1697 by the German chemist Ernst Stahl. But just what phlogiston was remained a matter of debate. For some it was an element itself. For others, it was more in the nature of a fiery essence, contained in combustible materials, without which combustion could not take place.

The concept of phlogiston gave rise to some anomalies. If it was a component of combustible materials, the residue ought to weigh less than the substances did before being burned. This was the case with substances like wood. But some metals, when heated, yielded a dull substance called a calx; in these cases the residue weighed more than the original metal. This anomaly was ignored by many advocates of phlogiston theory. Others rationalized it by suggesting that phlogiston had a *negative* weight, causing the residue to weigh more after the phlogiston had been given off.

Looking back now, when most educated people understand the role of oxygen in combustion, and every chemistry student knows that burning is a process of chemical change that results in no overall loss or gain of mass, it is easy to feel superior to these early searchers after truth. But they were able people; and their process of reasoning seemed to make sense in the light of the limited knowledge they possessed.

BENJAMIN FRANKLIN – SCIENTIST Benjamin Franklin, who is honoured as a founding father of the American nation, has an equal claim to be regarded as America's first scientist. He was born in Boston in 1706, the son of a candle-maker from Banbury, England, who had emigrated to escape religious persecution. The fifteenth of 17 children, he had only two years of formal schooling. When he was 10, his father took Benjamin into his business but, finding that he had no liking for it, and fearing he might run away to sea, apprenticed him to an older son, who was a printer. This gave Benjamin access to books; through them he contrived to give himself an education.

When he was 18, he was conned into travelling from Philadelphia to London, in the expectation of letters of introduction that did not materialize. However, he found work as a printer; while still in his teens, the books he printed, and his own writing, brought him into contact with some of the leading literary figures of the day. When he was 20, he returned to Philadelphia to work in a store owned by a friend. Soon afterwards, he re-entered the printing trade, and in 1730, when he was 24, he entered into a common-law marriage with a former girlfriend, Deborah, who had married while he was in London, but had been deserted by her husband. It was a union that would last until her death, 44 years later.

THE NATURE OF LIGHTNING Franklin had by this time become keenly interested in science; for the rest of his life, while engaged in writing, publishing, politics and diplomacy, he kept abreast of the latest developments, through contact with other scientists, and through his own experiments. In 1743 he founded America's first scientific society, the American Philosophic Society. He also found time to develop a number of notable inventions, including lightning conductors, bifocal lenses, and the Franklin stove.

Franklin was particularly interested in electricity and magnetism, which at that time were little understood. In 1745, a Dutch physicist named Pieter van Musschenbroek, who lived in the city of Leiden (or Leyden) invented an electric storage device that became known as a Leiden jar; it was Franklin's experience with this device that inspired his most famous experiment. Leiden jars, when touched, give off a spark and an electric shock. Suspecting that lightning was a form of electricity similar to the spark from a Leiden jar, Franklin decided that he would try to capture the electricity from a lightning bolt in one of his jars. One day in 1752, he fixed a wire spike to a kite, from which a silk thread led down to a key. He flew his kite towards a thundercloud and, when he placed his hand near the key, a spark leapt across the gap between them. He next succeeded in charging his jar from a lightning bolt, via the key, just as he might have charged it from a spark-generating machine. It was a thrilling demonstration that the lightning bolt and the humble spark from his jar were the same phenomenon. When he reported his experiment, it created a sensation, and it earned him membership of London's Royal Society. But he had

had a narrow escape. The next two people who tried the experiment were killed.

In the course of a long career he made many discoveries, and made important contributions to the developing study of electricity. In 1785, at the age of 79, he returned from Europe to Philadelphia, were he was elected president of Pennsylvania. He died in 1790, laden with scientific honours and degrees from universities in Europe and America. Twenty thousand people attended his funeral in Philadephia. He had put his 2 years of schooling to good use.

ANTOINE LAVOISIER From the time that Robert Boyle published *The Sceptical Chymist*, another century passed before chemistry acquired the language and the concepts that it needed to transform it into a respectable science. This transformation was assisted by the work of many able scientists, but one man stands head and shoulders above the rest. His name was Antoine Laurent Lavoisier, and it is no great exaggeration to call him, as some people do, the 'Newton of Chemistry'.

Lavoisier was born in Paris on 26 August 1743. His father was a well-to-do lawyer. He was himself intended for the law, and actually qualified to practise it but, as a result of hearing lectures by the astronomer Lacaille, he developed an enthusiasm for science. His early interest was in geology, and he did creditable work in that field. But he soon turned to chemistry, and this became a lifelong passion. In 1766, when he was only 23, he was awarded the Gold Medal of the French Academy of Sciences for an essay on the best means of lighting a large town.

Unlike some other scientists of his time – Cavendish, for example – Lavoisier was no shrinking violet. He lived a busy public life, and it was this involvement in public affairs that eventually proved his downfall. When he was 25 years of age, in 1768, he invested a large sum of money in the Ferme Générale, a privatized tax-collecting operation set up by the French government. Three years later, he married the 14-year-old daughter of one of the executives of the Ferme. It was an arranged marriage, but it was for many years a happy and productive one. His wife, Anne-Marie, was as intelligent as she was beautiful, and in their early years together they were never happier than when they were working in the laboratory together. As the years went by, and her husband began to spend more time away on business, Anne-Marie found consolation in the arms of one of his friends; nevertheless, they remained on good terms.

Lavoisier's intention when investing in the Ferme had been to acquire a reliable income to support him while he continued with his scientific researches. In this he was successful. The income, which derived mainly from taxes upon the poor, was enormous; and it enabled him to build and run a fine private laboratory – possibly the finest in the world – which became a meeting place for France's leading scientists, as well as for visiting celebrities such as Benjamin Franklin and Thomas Jefferson. In this way Lavoisier was able to keep up to date with the speculations, and the discoveries, of the leading scientists of the day. When he heard of a new idea, or an interesting experiment, he and Anne-Marie would at once embark on follow-up investigations of their own. He was not always as quick to acknowledge the work of others, however, or the contribution they made to his discoveries, and this led to several bitter disputes with fellow-scientists who believed their own work had not been properly acknowledged.

JOSEPH PRIESTLEY One of the men Lavoisier upset with his cavalier attitude was the English chemist and radical politician, Joseph Priestley. Priestley, who was 10 years older than Lavoisier, came from a very different background. He was born in Birstal, near Leeds, in 1733. He was the son of a Unitarian minister, and a minister himself; his religion had disqualified him from receiving a university education. Effectively self-taught after leaving school, he acquired a mastery of several languages, including Hebrew and Arabic. In 1766, when he was 33, he met Benjamin Franklin, who was in England as the representative of the American colonies, and it was this friendship that led him into a scientific career. Within a short time he had published a history of electrical research, and he followed this up with a history of optics.

The year after he met Franklin, Priestley was appointed minister of a chapel in Leeds. The chapel stood next door to a brewery, and Priestley became fascinated by the brewery processes. Fermenting grain gives off a gas, which we know as carbon dioxide. Priestley studied this gas, noting that it was heavier than air, and capable of extinguishing a flame. He dissolved it in water, and discovered that it gave the water a pleasant taste. He had discovered soda water; and he was rewarded for the discovery with the Royal Society's Copley Medal.

Priestley became particularly interested in gases, and went on to discover several more. When he began these researches, only three gases were known: air, carbon dioxide, and hydrogen (newly discovered by Cavendish, and named by Lavoisier). Priestley succeeded in isolating several more, including ammonia, nitrous oxide, and hydrogen chloride. In 1772, as a consequence of these discoveries, he was made a member of the

French Academy of Sciences, and obtained a comfortable appointment as companion and librarian to an English aristocrat, Lord Shelbourne. Two years later he made his most important discovery. He used a lens to heat the red substance known as mercury calx (mercuric oxide) in a tube. Metallic mercury was deposited, and a gas with some remarkable properties was produced in the upper part of the tube. When a lighted candle was placed in this gas, it burned much more brightly, and when a mouse was exposed to it, the mouse became particularly lively.

Unfortunately, Priestley was a devotee of the concept of phlogiston, and he was unable to appreciate properly the significance of his discovery. People who believed in the existence of phlogiston – the essence of heat – were aware that a candle in a sealed container soon goes out. They interpreted this as evidence that the air in the container had become saturated with the phlogiston from the burning candle, until it was unable to receive any more, and combustion ceased to be possible. Applying this reasoning, Priestley concluded that his gas was air that contained little or no phlogiston, and it was therefore 'hungry' for the phlogiston in the candle. He accordingly named his new gas 'dephlogisticated air'.

In October 1774, Priestley dined with Lavoisier in Paris, and acquainted him with his discovery. Lavoisier proceeded to conduct experiments of his own; after further correspondence with Priestley, he presented a paper to the Academy, in which he asserted that the key factor in combustion was Priestley's 'pure air', but without mentioning Priestley's name. Priestley was, understandably, very upset. Lavoisier had already established that when sulphur was burned it *gains* weight, instead of losing it.

During the next few years, he made a succession of ground-breaking discoveries, all of which were achieved as a result of his insistence on the importance of precise measurement. In 1779, he underlined his belief that Priestley's 'pure air' was not only a gas in its own right, but an element, by naming it *oxygen*. With the help of his fellow academician, Pierre Laplace, he conducted a series of experiments with live animals, as a result of which he was able to demonstrate that respiration was a form of combustion, wherein living creatures took oxygen from the air, in effect, to burn the 'fuel' they imbibed from their food.

In 1786 he published in the *Proceedings* of the Academy his dismissal of the will o' the wisp of phlogiston that had led chemists astray for so long. His list of 'bullet points' included two that really drove his message home:

1. There is true combustion ... only to the extent that the combustible body is surrounded by, and in contact with, oxygen; combustion cannot occur in any other kind of air, or in a vacuum, and burning bodies plunged into either are extinguished as surely as if they had been plunged into water.
2. In every combustion there is an increase in weight in the body burned; and this increase is exactly equal to the weight of the air that has been absorbed.

Even a scientist as great as Lavoisier could not be expected to throw off completely the system of thought in which he had been raised, and to the end of his days there was a residue of the old thinking in his writings. His theory of acids contained much that had to be put right later, as did his theory of heat.

But the chemists who came after him inherited a science utterly transformed by his work.

LAVOISIER'S CONTRIBUTION   Important though Lavoisier's discoveries were, they were only part of the contribution he made to the establishment of chemistry as a scientific discipline. Of equal importance was the lesson he taught: that sound conclusions could only be arrived at by carefully designed experiments and precise measurement. In his laboratory, it was the chemical balance that was the arbiter of scientific truth. He also gave chemistry a set of concepts that would prove immensely productive in the century that followed. It was he, rather than Boyle, who drew the distinction between an element and a compound in anything like a modern way. By doing so, he made it possible for chemists to begin to attach numbers to chemical processes. Thanks to these concepts, and the methods of precise analysis he pioneered, the nineteenth century became a chemical Golden Age.

It was a century he did not live to see. When the French revolution erupted in 1789, the hated tax farmers were obvious targets for the Terror that followed. And Lavoisier had had the additional misfortune of making an enemy of an aspiring scientist whom he had once treated with disdain. His name was Jean-Paul Marat, one of the most vigorous protagonists of the Terror. When the time came for settling scores, Lavoisier's scientific standing could not save him. On the morning of 8 May 1794, aged 53, and in his intellectual prime, he was tried and condemned to death. When he asked for a couple of weeks' stay of execution, in order that he might complete some scientific work, the judge replied, 'The Revolution has no need of scientists.' A few hours later, in what is now the Place de la Concorde, he went to the guillotine, with a calm and dignified air. One of his scientific contemporaries, the mathematician and astronomer Joseph-Louis Lagrange commented: 'It took them only an instant to cut off that head, but France may not produce another like it in a century.'

MEASURING LONGITUDE   Up to the middle of the eighteenth century, navigation at sea was a hit-and-miss affair. Such charts as existed were unreliable; and determining one's position in relation to the chart became more difficult the longer one was at sea.

In theory, position at sea can be fixed by plotting co-ordinates: lines that cross at the point of observation. One such line was relatively simple to plot. This was the line of latitude. The latitude of a ship or a rock is its distance in degrees north or south of the equator. This is easily determined on a sunny day by using a sextant to measure the angle between the Sun and the horizon at midday, and adjusting this by the corresponding angle at the equator on the day in question (from printed tables).

Unfortunately, latitude, by itself, is not much use. A ship's captain in the Atlantic who was told he was 42 degrees north of the equator knew only that he was somewhere between Cape Finisterre and Cape Cod. What he needed was both latitude and longitude. Longitude is the distance, measured in degrees, east or west of a reference location on the Earth's surface. Without a scientific method of establishing longitude, sailors had to rely on 'dead reckoning': a process of guessing how far one had travelled in a day, and in what direction, and marking this distance and direction on the chart. One day after leaving harbour, this was a reliable method of plotting position, but after 20 days on the open ocean, it was rather less accurate.

The problem of determining longitude had engaged the attention of some of the best minds in Europe. Galileo, Huygens and Newton had tackled it, but had failed to come up with an answer. In 1714, the British government offered a prize of £20,000 (£1 million in today's money) to anyone who could come up with an acceptable solution.

Among the people who competed for the prize was the British astronomer royal, Nevil Maskelyne, who felt sure that the solution was to be found in what was called the 'lunar distance' method. But the man who solved the problem was neither a mathematician nor an astronomer. He was a clockmaker named John Harrison.

TIME IS DISTANCE The reason a clockmaker was able to make the seas safe for sailors was a simple relationship between time and distance. The Earth revolves on its axis once every 24 hours. Every line of latitude is a circle of 360 degrees, and so is the Sun's path around the sky. The Sun traverses the entire 360 degrees of the sky once every 24 hours. This means that it travels 15 degrees farther west every hour. The Sun is directly overhead at noon, local time. After an hour has passed, the Sun will be vertically overhead at a place (call it B) 15 degrees west of the place (call it A) where it was an hour earlier. If, when it is noon at B, we know that it is 1.00 p.m. at A, then we know we are 15 degrees west of A.

It was not difficult to establish when the Sun was at its maximum height; this was, by definition, noon local time. The problem was to know exactly what the time was at a particular reference point, such as the Royal Observatory at Greenwich. In order to know this, a ship would have to carry chronometers set permanently to Greenwich time, and these would need to keep exact time for months at a stretch, in any climate, and no matter how rough the seas. This was the challenge that Harrison faced. Eventually, in 1759, after nearly half a century of inspired craftsmanship, he produced a chronometer that met these exacting requirements.

Thanks to Harrison's chronometers, it became possible for ships' captains to fix their position precisely, and surveyors to plot accurate charts. But Maskelyne saw to it that Harrison didn't get the prize, which was never won. It took an appeal to the King — and an Act of Parliament — to get Harrison a reward for the service he had rendered his country, and the world.

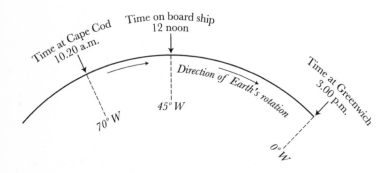

Figure 11. The Relationship between Time and Longitude
If it is 3.00 p.m. in Greenwich when it is noon on board ship, the ship must be 3 × 15° = 45° west of Greenwich.

WEIGHING THE EARTH (ACT ONE)   Maskelyne's plotting against Harrison is remembered now to his disadvantage. But in his capacity as astronomer royal he was involved in a scientific enterprise that puts him in a better light.

A century after the publication of Newton's *Principia,* there was an experiment suggested in the book that had still not been performed. Although Newton had established the existence of the force of gravity, he had not been able to put a value on it. Calculations involving gravity were all based on the *relative* attraction between objects of differing masses. The value that would enable one to ascertain the *absolute* strength of gravity in any given situation – the *gravitational constant* – was unknown. Newton suggested that if a plumb line were suspended in the vicinity of a mountain, the gravitational attraction of the mountain should pull it slightly away from the vertical, and the deflection might be large enough to be measurable. If it were, the amount of the deflection would enable one to calculate the relative masses of the Earth and the mountain. If the mass of the mountain could be estimated with reasonable accuracy, it would then be possible to arrive at a value for the mass of the Earth, and therefore for the gravitational constant. Since the Earth's volume was known, this would also make it possible to ascertain the average density of the Earth.

The problem was to find a suitable mountain. In order to estimate its mass, its average density would have to be guessed at. If its volume had to be guessed at as well, any errors would multiply to an unacceptable degree. So the mountain needed to have a regular shape, to make its volume reasonably certain. At Maskelyne's prompting, the Royal Society launched a search for a suitable mountain. The task was assigned to a surveyor friend of Maskelyne's named Charles Mason (the Mason of the Mason–Dixon Line), who reported back that he had found a beautifully proportioned mountain called Schiehallion in the Scottish highlands, which looked ideal for the purpose. The survey of the mountain was supervized by Maskelyne himself, who in 1774 spent four months in camp at its foot. The calculations were delegated to a young mathematician named Charles Hutton, who came up with the first-ever scientifically calculated figure for the mass of the Earth: 5×10^{21} tons (5,000,000,000,000,000,000,000 tons).

The calculation of the Earth's mass was an exciting event in its own right, but the significance of the result was immensely greater than this. Because Newton's theory had already established the *relative* masses of the Earth, the Sun, the Moon, and all the planets, it was now possible to calculate their *actual* masses. A mere 165 years after Galileo had first pointed his telescope at the skies, the entire solar system had been weighed and measured. Astronomy had truly come of age.

WEIGHING THE EARTH (ACT TWO)   Maskelyne was justifiably pleased with the outcome of his campaign to establish the Earth's mass, but others were less satisfied. The calculation had only been possible on the basis of guesswork. The neat shape of the mountain, and the careful measurement of its dimensions, provided some assurance that the estimate of its volume could be relied upon. But to arrive at the key figure in the calculation – the mountain's mass – it was necessary to estimate its density. This was done by making assumptions about the rocks of which it was composed. If these assumptions were incorrect, the answer would be incorrect. Hutton's estimate of

5,000 million, million, million tons was a useful approximation; but people were soon looking for a way of arriving at a more accurate figure. In 1798, another Englishman obtained the precise measurement the world was looking for – and did so without leaving his house.

Henry Cavendish was born in 1731 in the French city of Nice, where his mother was living for health reasons. She died when he was two. He was educated in England, and spent four years at Cambridge University; but he did not submit himself for a degree, being too shy to face the examiners. He was the grandson of two dukes; and he inherited a fortune from an aunt, which made him one of the richest men of his time. He was also one of the most reclusive. He lived alone, avoided visitors, and ordered his meals by leaving a note for his housekeeper. He lived simply, and had no interest in money. His bank manager once commented on the fact that he had the equivalent of about £5 million in today's money in his current account, and suggested that he might consider placing it where it could earn interest. Cavendish told him that if he ever bothered him again, he would move his money elsewhere.

Cavendish inherited his interest in science from his father; and it was his passion for 60 years. He cared nothing for fame, and published little, with the result that many of his discoveries remained unknown until after his death. His name is commemorated in the Cavendish Laboratory in Cambridge, and in the famous 'Cavendish Experiment', which he conducted. This was devised by his friend John Michell, a clergyman and a keen geologist. Michell had designed the apparatus himself, but he had died before he could perform the experiment. Cavendish acquired his equipment, and had it re-erected in one of his London houses.

The apparatus was simple. It consisted of 2 metal balls, 30 centimetres/12 inches in diameter, suspended from a steel gantry; and 2 smaller balls, 5 centimetres/2 inches in diameter, separately suspended close to them, connected to one another by a fine copper wire. It constituted what is technically known as a torsion balance. It was designed to measure the twisting movement created in the wire by the gravitational attraction of the larger balls upon the smaller, as they were moved about by pulleys operating on the beam from which they were suspended.

So as not to disturb the equipment, the experiment was conducted by remote control. Cavendish used a telescope, mounted outside the room, to read the minutely graduated scale measuring the movement (in hundredths of an inch), which was itself illuminated by a narrow beam of light directed from outside the room.

Gravity is a weak force, and the measurements Cavendish was proposing to make were so fine as to almost defy belief. But he and his apparatus were equal to the task; and he came up with a density of the Earth of 5.48 times the density of water. This was 20 per cent greater than the figure arrived at in the Schiehallion experiment, and within 1 per cent of the figure accepted today. After his death, it was discovered that he had made an error in his calculations, without which his result would have been 1.5 per cent adrift of the correct value. But given that the attraction of the balls upon one another was only a $\frac{1}{50,000,000}$th part of that exercised upon them by the Earth, he can be forgiven the inaccuracy.

MICHELL'S SPECULATIONS John Michell, who designed the experiment that now bears his friend Cavendish's name, was a man of remarkable gifts. Before taking up his appointment as rector of the Yorkshire parish of Thornhill, he had held the post of professor of geology in the university of Cambridge. After moving to Yorkshire, he maintained his interest in science, and his speculations extended far beyond the earth sciences that had had been his official speciality. His most striking speculation was contained in a paper that Cavendish read on his behalf to a meeting of the Royal Society in 1783. In this paper, Michell considered the implications of the finite speed of light, in the context of Newton's gravitational theory.

Every celestial body has an *escape velocity*, the velocity that must be attained for an object to escape from that body's gravitational influence (*see page 79*). For any given distance from the centre of gravity in question, the escape velocity is directly proportionate to the mass of the parent body. If light itself is subject to gravitational attraction, it is theoretically possible that a body might exist whose mass is so great that its escape velocity would exceed the speed of light. Such a body would by definition be invisible, since its light could not escape. Michell not only suggested the possibility of such a body, he worked out how big it might have to be. Assuming it had a density similar to that of the Sun, he calculated that, if its diameter were more than 500 times that of the Sun, its escape velocity would exceed the speed of light, and it would be invisible. In modern parlance, it would be a 'black hole'.

Michell's paper went on to consider how, if we could not see a such a body, we could know it existed. The answer he gave was again rooted in Newton's theory. If a black hole had a visible companion in orbit around it, it would be possible, from the companion's motion, not only to deduce the black hole's existence, but also to calculate its mass. It is a striking thought that 200 years before the reality of black holes was generally accepted, a Yorkshire clergyman was positing their existence, and suggesting how they might be weighed.

In 1755 the Portuguese city of Lisbon was laid waste by one of the most destructive earthquakes of modern times. Michell suggested that the earthquake had originated under the sea, and went on to speculate that the epicentre of such earthquakes might be determined by noting the time at which tremors were felt in various locations. This became standard practice in the twentieth century, and Michell is now recognized as the 'Father of Seismology'.

HOW FAR TO THE STARS? In 1784, Michell presented an argument in favour of the proposition that the stars were light years distant from the Earth. It would be another 54 years before the German astronomer Friedrich Wilhelm Bessel would prove the truth of Michell's assertion by making the first-ever measurement of the distance to a star.

LINNAEUS AND HIS SYSTEM Some of the most important advances in science have come as a result of rearranging facts in a different order. The rearrangement that paved the way for two centuries of advances in biology was the work of a Swedish doctor, Karl von Linné, who is known to science by his Latin name of Linnaeus.

Linnaeus was born in Råshult, in southern Sweden, in 1707. From his earliest years he had a love of plants, which he shared with his father, the village pastor. He studied medicine at Uppsala, but by the age of 25 he was already teaching botany there. In 1732 he led a university expedition to Lapland, in the course of which he collected specimens of hundreds of hitherto unknown plants.

In 1735, following travels in England and elsewhere in western Europe, he published the book that ensured his undying fame. It was called *The System of Nature*, and it proposed a way of classifying plants and animals quite different to anything that had gone before. Many other naturalists, from Aristotle onwards, had attempted to classify life forms; but all previous systems had been based to a greater or lesser extent on superficial features such as flower colour, or on behavioural characteristics, such as 'creatures that swim'. Linnaeus based his classification on shared characteristics of a fundamental nature. So, for example, mice and whales were grouped together as mammals, and flowering plants were grouped separately from non-flowering plants. An important feature of his system was the concept of 'nesting', whereby large groups were subdivided into more narrowly defined categories. Thus vertebrate animals were subdivided into mammals, reptiles, birds, etc, and mammals were further subdivided into carnivores, insectivores, and so on.

Linnaeus' system embraced the whole living world, but it was rooted in his study of plants, and it was through his study of plants that he developed the preoccupation with sex that permeates all his work. Some passages from his descriptions of the physical parts of plants, and their methods of reproduction, read more like extracts from bodice-rippers than from scientific treatises. But it was a preoccupation that had good scientific reasons behind it: it enabled him to transform the study of botany, and effectively invent the modern science of biology.

Apart from the successive division into groups and subgroups, the other key feature of Linnaeus' approach was his adoption of a *binomial system*, whereby every species was identified by its own specific Latin name and the Latin name of the immediate group to which the species belonged. Thus lions were *Felis leo*, and wildcats were *Felis sylvestris*, both belonging to the genus *Felis* (cats). This is still the basis of biological nomenclature, and it remains the privilege of anyone who first describes a new species to decide the double-barrelled name by which it is known.

The first edition of *The System of Nature* was a 7-page pamphlet. By the tenth edition, Linnaeus' passion for classification (or taxonomy, to give it its proper name) had turned it into a book 2,500 pages long. His criteria for dividing up the living world were more fundamental than those employed by his predecessors, and his system has stood the test of time. But, by the standards of those who came later, he too could be guilty – albeit to a lesser extent – of being deceived by superficial differences, and his system has since been refined.

When later biologists revised his categories, rearranging them to emphasize family relationships, this brought out an irony contained within his work. Linnaeus did not employ the 'family' category himself, and he did not believe in evolution. Although he later became less rigid in his belief in the fixity of species, he never wavered from his belief that every genus of animal and plant represented an original creation, and embodied a separate idea in the mind of God. But the 'nesting' aspect of his system lent itself to illustration in the form of family trees, and family trees inevitably suggested shared parentage. By the time he died in 1778, his 40 years' devotion to the cause of taxonomy had laid the foundations on which Charles Darwin would construct his theory of evolution by natural selection.

CLASSIFYING SPECIES Linnaeus' system of classification recognized only four categories: class, order, genus, and species. Believing as he did in the persistence of species from their original creation, his system was designed as an aid to the identification of species, not the construction of family trees. He therefore had no use for higher categories such as phyla or kingdoms. Linnaeus himself, classified according to the present-day version of his system, would be described as follows:

Kingdom:	Animalia
Phylum:	Chordata
Subphylum:	Vertebrata
Class:	Mammalia
Order:	Primates
Family:	Hominidae
Genus:	*Homo*
Species:	*Homo sapiens*

This tells us he was an animal, with a spinal cord and a spine; that he was a mammal; that he belonged to the same order as the apes and monkeys, and the same family as extinct humanoids such as *Homo erectus*; and that he was a member of the same species as every human who has walked the Earth in the past 100,000 years.

THE AMAZING DIVERSITY OF LIFE The system that Linnaeus invented has had to cope with a diversity far beyond anything he could have imagined. Since his death, biologists have described over a million different species; and there can be no doubt that a greater number awaits discovery. And that is just *living* species. Extinct species would increase the total several times over. The British biologist J.B.S. Haldane was once asked what his study of nature had taught him about the Almighty. He replied that He seemed to have 'an inordinate fondness for beetles'. It was an answer he would have no reason to reconsider today. Of the 750,000 recognized species of insects, 330,000 are beetles.

The current list of living species registered by the World Resources Institute divides up as follows:

	Number of species* (rounded off)
Vertebrates	
Mammals	4,000
Birds	9,000
Reptiles	6,000
Amphibians	4,000
Fish	19,000
Invertebrates	
Arthropods (spiders, crabs, etc., incl. 750,000 insects)	870,000
Echinodermata (sea urchins)	6,000
Molluscs (shellfish)	50,000
Annelida (segmented worms)	12,000
Platyhelminthes (flatworms)	12,000
Nematoda (roundworms)	12,000

	Number of species[*] (rounded off)
Coelenterates (jellyfish, corals)	9,000
Porifera (sponges)	5,000
Plants	250,000
Fungi	7,000
Protists (algae, etc)	8,000
Prokaryotes (bacteria)	5,000
Archaea (heat-loving, etc. unicellular forms)	Not listed

[*]About 30,000 new species are identified every year

THE GENUS HOMO Although Linnaeus, according to his own system, belongs in the genus *Homo*, separate from all the other primates, it was a separation he himself would have preferred to avoid. In the foreword to his book *The Animals of Sweden,* which was published in 1746, he said, 'I have yet to find any characteristics that enable man to be distinguished on scientific principles from an ape.'

His nerve, alas, failed him in the end. Like many respectable naturalists in the eighteenth and nineteenth centuries, he allowed his science to be influenced by religious dogma. In the final version of his classification, he placed *Homo sapiens* in a genus of its own. It was an honour the species did not merit. If we were designing the classification today, it is an honour that would not be extended.

THE GREGORIAN CALENDAR In 46 BC, the dictator Julius Caesar, who was married to a Roman lady of good family, upset some of the most influential people in Rome by extending a lavish welcome to his lover, Queen Cleopatra of

Egypt. He gave her use of a villa for the duration of her visit, which lasted until his murder two years later; it undoubtedly encouraged the conspiracy that brought about his death. But a more long-lasting consequence of his Egyptian dalliance was his reform of the calendar, which was recommended to him by an Egyptian astronomer named Sosigenes, who accompanied him back to Rome.

The Julian calendar, which Caesar introduced and which bore his name, remained in force throughout most of Europe for 1,600 years after his death. Unfortunately, it was not as accurate as it might have been, and by the eighth century AD it was starting to cause disquiet, because it was leading to problems in the fixing of the Christian festival of Easter. For the next 800 years, the subject of calendar reform was one of the most intensively debated issues in Christendom.

The problem was that the Julian calendar assumed that the length of the year was 365.25 days, whereas the true figure is 365.242 days. So by the late sixteenth century, the Christian calendar was badly adrift. For reasons we need not go into here, the accumulated error was 10 days. In 1582 Pope Gregory XIII persuaded a number of European states to agree to what became known as the Gregorian calendar. To enable the old calendar to catch up, 10 days were dropped altogether.

The Julian calendar had achieved an average of 365.25 days by slipping in an extra day once every 4 years. The Gregorian made a slight change to this rule: an extra day was to be inserted if the year was divisible by 4, but *not* if it was divisible by 100 – unless it was also divisible by 400.

Most European countries adopted the new calendar fairly quickly. The English bishops, reluctant to follow a Popish lead, urged further discussion which, in very English fashion, continued for 170 years. By the time the English (and the Scots and Irish) came in line, they needed to drop 11 days. Wednesday 2 September 1752 was followed by Thursday 14 September. Not everyone was happy. Believing they had been robbed of a piece of their lives, protesters adopted the slogan 'Give us back our eleven days'. But once the deed was done, the expansion of the British Empire ensured that the calendar was adopted around the world. The Russian Empire, however, held out until after the 1917 Revolution.

CHANGING THE CALENDAR Several countries adopted the Gregorian calendar almost immediately; but as the following table shows, some took a long time to make the change:

Date of adoption

Italy, Spain, Portugal, Poland	1582	5–14 October
France	1582	10–19 December
Germany (Catholic)	1583	various dates
Germany (Protestant)	1700	19–28 February
England (and colonies), Scotland, Ireland	1752	3–13 September
Alaska (previously part of Russia)	1867	
Japan	1873*	
China	1912*	
Soviet Union	1918	1–13 February
Greece	1924	10–22 March
Turkey	1926	19–31 December

* These countries did not use the Julian calendar.

COWPOX AND SMALLPOX In the eighteenth century, smallpox was a dreaded disease. In some outbreaks, it killed one in three of those who contracted it, and those who survived were often left blind or horribly disfigured. It had been known for a long time that people who had suffered a mild attack of the disease were thereafter immune to it. Both the Turks and the Chinese had discovered that it was possible to protect people by inoculating them with pus taken from the sores of smallpox sufferers, and the practice had been introduced into Europe. But it was far from satisfactory. Although the severity of an attack was often reduced by prior inoculation, the procedure also spread the disease to the previously uninfected. The man who discovered how to apply inoculation safely to the treatment of smallpox was an English doctor named Edward Jenner.

Jenner was born in Gloucestershire in 1749. His father was a clergyman, but his parents died when he was a child, and he was taken care of by an older brother. When he was 13, he was apprenticed to a surgeon; and at 21 he went to London as a pupil of the anatomist John Hunter, who was the most eminent physician in England. Master and pupil shared not only a trade, but also an interest in all branches of science, in particular in natural history. Jenner became a favourite in London society, and he was engaged by Sir Joseph Banks to catalogue the geological specimens brought back from Captain James Cook's first voyage.

Returning to Gloucestershire as a doctor, Jenner became interested in the subject of cowpox. Dairymaids were noted for their clear complexions; and many country people attributed this to their having avoided smallpox as a result of having been infected with cowpox. This led Jenner to consider the possibility of inoculating with cowpox as a protection against smallpox. Other people had attempted this with unfortunate results, but Jenner's researches revealed that there were two kinds of cowpox, only one of which appeared to confer immunity from smallpox. He also became persuaded that, for immunity to be obtained, cowpox had to be contracted at just the right point in the disease's progress.

Cowpox was uncommon in Gloucestershire, and it was not until 1796 that an opportunity arose for the experiment he had in mind. In that year, he came across a dairy maid named Sarah Nelmes, who had caught cowpox from her cow Blossom. Using a penknife, and pus from Sarah's sores, he inoculated an 8-year-old boy who had had neither cowpox nor smallpox. At great risk to the child, and no little risk to his own reputation, he not only inoculated the boy with smallpox, but subsequently exposed him to the disease on a number of occasions. No adverse consequences followed, and in 1798 he repeated the experiment, again with complete success. He now felt safe in publishing details of his experiments. He became a celebrated figure, and received two substantial money prizes from a grateful nation. After some opposition, a programme of inoculation was embarked upon. Millions of people were vaccinated (a word Jenner coined from the Latin word for a cow), and within a few years, deaths from smallpox had fallen by two-thirds.

The last recorded case of smallpox in the world occurred in 1978, and soon afterwards the World Health Organization declared that the disease had been completely eradicated. Thanks to Jenner's scientific curiosity, and his gambler's lucky throw, one of humanity's greatest scourges had been banished from the Earth.

JOHN GOODRICKE The giants of science – Newton, Faraday, Einstein, and the like – tend to occupy the limelight, crowding out others; so let's hear it for John Goodricke. He died, aged 21, in 1786, with a remarkable achievement to his name and a promise never to be fulfilled.

Goodricke, the son of an English diplomat and his Dutch wife, was born in Groningen, in the Netherlands, in 1764. When he was five, an attack of scarlet fever left him stone deaf; but his parents sent him to a special school in Edinburgh, where he learned to lip-read and develop his speaking ability. While attending another special school near York, he became a keen amateur astronomer; in 1782, when he was 18 years of age, he made an inspired guess about the secret of a star.

The second brightest star in the constellation of Perseus is called Algol, an Arabic name meaning 'the ghoul'. Whoever named it had noticed something strange about it. It is what astronomers call a *variable* star. Once every three days, its light suddenly diminishes by three-quarters; and it remains dim for ten hours, before resuming its original brightness. The Greeks had nothing to say about it. Perhaps it was an embarrassment in what was supposed to be a perfect and unchanging universe. European astronomers had known of it since at least 1670, but it remained a mystery.

Goodricke studied the star closely, and discovered that the fluctuations in its brightness occurred with absolute regularity over a period of 68 hours, 49 minutes. To explain it, he ventured a daring speculation. His suggestion was that Algol had an invisible companion in orbit around it; and that the diminution in brightness occurred when this companion passed between Algol and the Earth, cutting off most of its light. In the spring of 1783, when he was still only 18, Goodricke presented a paper outlining his speculations to the Royal Society. In April 1786, at the age of 21, he was admitted to membership of the Society, and awarded its coveted Copley Medal. Four days later, he died of pneumonia.

In 1890, 104 years after Goodricke's death, a German astronomer, Hermann Vogel, trained his spectroscope on Algol, and established from its spectrum that Goodricke's explanation had been correct. Algol was an *eclipsing binary*: a bright star that was periodically obscured by a dark companion circling round it. In a just world, it would be called Goodricke's star.

LIGHT AND SOUND   Light and sound have much in common. Both convey information about distant events to organs that have evolved to handle a particular kind of message. Both display a wave-like character; and both have a spectrum. Both are distorted by the passage from one medium to another. And both are dispersed in accordance with Newton's inverse square rule. A man sitting 4 feet/1 metre away from the TV may be exasperated by his wife's complaints that she is missing all the jokes. But if she is sitting 6 feet/1.5 metres away, the volume of sound reaching her ears will be less than half that reaching his, and her complaint may be well founded.

But there are some fundamental differences between the two phenomena. Light is like a message fastened to an arrow. Sound is like a message carried by a team of relay runners. It is a disturbance that spreads through a material as a result of collisions between the molecules of which the material is composed. If there are no molecules to pass it on, the message can't get through. That is why sound cannot travel through a vacuum, whereas light loves one.

Light travels much faster than sound. The speed of light in a vacuum is a stunning 300,000 kilometres/186,300 miles per second. The speed of sound in air is a mere 340 metres/1,100 feet per second. This makes it easy to track the path of a passing thunderstorm by counting the interval between the lightning flashes and the rolls of thunder associated with them. If the time lag is 5 seconds, the lightning is 1.7 kilometres/1 mile away.

THE CHANGING SPEED OF LIGHT AND SOUND Both light and sound travel at varying speeds as they move from one medium to another. Specimen figures for light are as follows:

Medium	Speed		Refractive index*
	(km per second)	(miles per second)	
Vacuum	300,000	186,300	1.00
Air	299,500	186,000	1.00
Water	225,000	140,000	1.33
Glass (average)	185,000	115,000	1.60

*The amount by which light is bent by its passage through a particular material.

Unlike light, sound travels faster through dense materials:

Medium	Speed	
	(metres per second)	(ft per second)
Vacuum	nil	nil
Air (at 0°C/32°F)	330	1,090

Medium	Speed	
	(metres per second)	(ft per second)
Water (at 0°C/32°F)	1,280	4,210
Wood (oak)	3,850	12,360
Steel	5,060	16,600

THE AIR WE BREATHE The composition of the lower atmosphere (up to about 15 kilometres/10 miles high) is remarkably consistent around the globe. It is basically four parts nitrogen to one part oxygen. The make-up of dry air (containing no water vapour) is:

Nitrogen	78.1%
Oxygen	20.9%
Argon	0.9%
Carbon dioxide	0.03%★
Other	0.07%

★ The proportion of carbon dioxide in the air is lower in coastal areas and around the poles. In urban and industrial areas it also varies according to the scale of fossil fuel consumption.

AIR AT ALTITUDE Although the composition of the air is much the same throughout the lower atmosphere, the actual quantity of air falls off sharply as one rises above the ground. The weight of the atmosphere itself compresses the air at lower levels, making it denser. There is thus much less air contained in a given volume at higher levels. The following table illustrates the rapid fall in the density of the air with increasing height:

Height (metres/ft)	Density of the air
0	100%
1,000/3,200	90%
2,000/6,400	80%
4,000/12,800	67%
8,000/25,600	43%

Eight thousand metres is the height of the highest peaks of the Himalaya. It might be thought that since the air at this height is about half as dense as the air at sea level, and contains the same proportion of oxygen, survival at this height would be a simple matter of breathing twice as fast. But the physiology of breathing and oxygen circulation is a complex matter, and an non-acclimatized mountaineer – forced to breathe the air at 8,000 metres/25,600 feet – would not last long.

KEEPING WARM Another lesson mountaineers learn the hard way is the importance of wrapping up when it is windy. Even if the outside temperature is low, it is usually possible to stay warm, if the right clothing is used, and the air is still. It is a different problem altogether if one is confronted with low temperature and high winds. It is in such circumstances that *wind chill* comes into play, and wind chill is a killer.

In still air, the body's warmth creates an insulating blanket of not-so-cold air around it. This lessens the temperature gradient between the skin and the surrounding air, and reduces the rate at which heat is lost. But any flow of air at a lower temperature than the temperature of the skin increases the rate of heat loss. Contrary to what might be expected, the effect is particularly marked at moderate wind speeds. This is because a 20-mile-an-hour wind moves across the skin at twice the speed of a 10-mile-an-hour wind, whereas a 60-mile-an-hour wind moves across the skin at only 1.5 times the speed of a 40-mile-an-hour wind.

Meteorologists employ *wind-chill factor* to indicate the effective temperature created by a combination of wind and low temperature. The life-threatening nature of even modest wind speeds in cold conditions is conveyed by the following tables:

EFFECTIVE TEMPERATURE CREATED BY WIND CHILL

Wind speed (km per hour)	Ambient temperature (°C)			
	5	–10	–20	–30
10	2	–12	–28	–40
50	–2	–17	–35	–48
100	–4	–20	–40	–54

Wind speed (miles per hour)	Ambient temperature (°F)			
	40	20	0	–20
5	36	13	–11	–34
30	28	1	–26	–53
60	25	–4	–33	–62

WIND SPEED Mountains can be windy places. The highest wind speed ever recorded was on Mount Washington, New

Hampshire, on 12 April 1934. It registered 371 kilometres/ 231 miles per hour. That's *three times* hurricane force.

PLANETARY ATMOSPHERES   The fact that we have any air to breathe at all is down to gravity. The concept of *escape velocity* (*see page 79*), which governs the ability of a spaceship to leave the Earth, governs the behaviour of anything else which might otherwise take off into space, gases included. The velocities of the individual molecules in a given gas vary, but each gas has a characteristic average molecular velocity. Roughly speaking, the lighter the gas, the faster the average speed of its molecules. The composition of a planet's atmosphere is therefore dependent on two factors:

1. the quantity of various gases generated at the planets surface up to the present time,
and
2. the escape velocity at various levels of the planet's atmosphere.

In the case of the Earth, the average velocity of the molecules of oxygen and nitrogen in the atmosphere is significantly below the escape velocity of 11 kilometres/7 miles per second, so only a tiny proportion of these gases, from the topmost layer of the atmosphere, seeps out into space in any given period. It therefore requires only a modest production of these gases at the Earth's surface to keep the proportion of these gases in the atmosphere in equilibrium. Hydrogen and helium, being much lighter, have much higher molecular velocities, and they are therefore found in only very small quantities. This is in marked contrast to the massive planets Jupiter and Saturn, whose atmospheres are mainly composed of these elements.

If the molecular velocity of a gas is significantly above the escape velocity of the planet in question, that gas will rapidly disappear into space. It follows that a planet with a low escape velocity is unlikely to have any lighter gases in its atmosphere, and may well have no atmosphere at all. It is for this reason that there is virtually no atmosphere on Mars, and none whatsoever on the Moon. Such atmosphere as it may once have possessed has long since disappeared into space.

THE UPPER ATMOSPHERE The thin air at the top of Everest is still, scientifically speaking, part of the atmosphere's lower depths. Scientists call the region up to about 16 kilometres/10 miles high, or roughly twice the height of Everest, the *troposphere.* This is where the clouds are, and where the weather action is.

The next layer up is the *stratosphere.* This extends from about 16 kilometres/10 miles up to about 50 kilometres/30 miles above the surface. As one goes up through the stratosphere, the temperature increases somewhat, as a result of the heat released by the process in which a variety of oxygen called ozone is generated by the action of ultraviolet light.

Above the stratosphere, from 50 kilometres/30 miles up to about 80 kilometres/50 miles, is the *mesosphere,* where the temperature cools with increasing height. At this level the atmosphere really is extraordinarily thin, but there are still enough molecules around to offer resistance to a fast-moving object; and it is at this level that most meteors light up – and burn out. It is at this level too that ice crystals form the highest clouds of all, *noctilucent* clouds, which are seen after dark, lit up by the invisible Sun, below the horizon.

Above the mesosphere are the thin layers that make up the *ionosphere*. It is from these layers that long radio waves are reflected, making it possible for listeners on the ground to receive AM transmissions from beyond the optical horizon. Above these layers is the *thermosphere,* which extends to about 500 kilometres/300 miles. Beyond the thermosphere, from 500 kilometres/300 miles up to several thousand kilometres, is the *exosphere*. The atmosphere there is so thin that on Earth it would be called a vacuum. This is a region of magnetic fields and high-energy particles; space stations stay well below it. From the point of view of physicists, it is still part of the Earth's atmosphere; but for the rest of us it is just 'space'.

DALTON AND THE ATOM We are so used to the concept of atoms as the basic building blocks of everything in the universe – ourselves included – that it is difficult to imagine a conversation in which their existence is not taken for granted. But a little over 200 years ago, the mere hint of a belief in such entities might well have been regarded as fantastic. The man who made atoms respectable was an English chemist named John Dalton.

Dalton was born in the village of Eaglesfield, in the county of Cumberland, in 1766. He was a Quaker – the son of a weaver – and he left school at the age of 11, to be a teacher in a Quaker school. His first scientific interest was meteorology, which he began to study in his early twenties, with instruments of his own design. In 1793 he published a book entitled *Meteorological Observations and Essays,* which was one of the first of its kind ever written. He maintained his interest in the subject for over 50 years, making weather observations – 200,000 in all – until the day he died. In 1794, he became the first person to describe colour-blindness: a subject he was well qualified to write about, since he had the condition himself.

It was a small step from thinking about the weather to thinking about air; and from thinking about air to thinking about the properties of gases in general. He soon came to believe that all gases were composed of minute, invisible particles, and before long he had come to a similar conclusion about liquids and solids as well.

The French chemist Joseph Louis Proust had demonstrated in 1799 that copper carbonate contained the elements copper, carbon, and oxygen in the proportions 5:1:4; and that these ratios applied whether the compound was obtained in the laboratory, or from nature. He later established that the same was true of other compounds, and he went on to embody this principle in a statement that became known as the *law of definite proportions.*

Dalton realized that such a law would be an inevitable outcome if:

1. elements were composed of minute particles;
2. the particles of one element all had the same mass;
3. the particles of different elements had different masses; *and*
4. the combination of elements took place at the level of the individual particle.

He further surmised that elements combined in varying proportions to form different compounds. This led him to suspect that methane and ethylene were compounds in which nitrogen and carbon combined in different ratios; and that the same was true of carbon monoxide and carbon dioxide. He embodied this conclusion in a refinement of Proust's law, which he called the *law of multiple proportions.*

Dalton realized that his particles resembled the *atoms* that the Greek philosopher Democritus had long ago suggested were the building blocks of nature. He accordingly adopted Democritus' word to describe his own particles. But he owed no debt to his Greek predecessor. Democritus' atom was a purely philosophical concept, without theoretical or experimental support.

In 1808, Dalton set out his ideas in a book entitled *New System of Chemical Philosophy.* Within a few years, his concept of chemical combination as the joining together of atoms of the various elements, in fixed proportions, in a process that was susceptible to precise measurement, became the shared view of most practising chemists. It was beginning to look as though chemistry might at long last be about to take its place among the exact sciences.

Dalton's modest nature, and his Quaker beliefs, prevented him from accepting many of the honours that governments and learned societies wished to lavish upon him. But he did agree to accept an honorary degree from the University of Oxford. His ceremonial gown was scarlet – a colour a Quaker was not supposed to wear. But his colour-blindness came to his rescue. In his eyes, his gown was grey.

THE GALVINIC PILE In 1791, Luigi Galvani, lecturer in anatomy and professor of obstetrics in the University of Bologna, published a paper in which he described an experiment involving frogs. He recounted how, when a frog had been laid out for dissection on a table on which there stood an electrical machine, the frog's legs had begun to twitch. He also described how the legs twitched when they were laid out on a metal surface during a thunderstorm, or when they were simultaneously touched by instruments made of different metals. He concluded that he had witnessed the release of 'animal electricity', electricity stored within the frog's body.

Galvani sent a copy of his paper to his friend Alessandro Volta, professor of physics at the University of Pavia, hoping for his support. But the friendship was put under strain when Volta insisted, in papers published in 1792 and 1793, that the twitching was caused by an external electric current, and that, in the case of the instruments, the electricity had been generated by a reaction between the two metals. To support his theory, Volta experimented with various combinations of metals to see whether they were capable of generating an electric current. He placed his tongue on the terminals to assess the strength of the current generated, and this suggested to him that the saliva in his mouth was contributing to the effect produced. He therefore experimented with plates of various metals in assorted liquids. In 1800 he created the 'voltaic pile', the world's first working wet battery, consisting of alternating discs of silver and zinc interspersed with brine-soaked layers of cardboard.

In 1801 Volta was invited to Paris to demonstrate his device to Napoleon, who was so impressed that he made him a count, and a member of the Legion of Honour.

ELECTROCHEMISTRY In March 1800, Volta sent a letter to Sir Joseph Banks, president of the Royal Society, with a sketch of his new invention. News of the letter reached the ears of a waterworks engineer turned popular science writer called William Nicholson, who promptly set about constructing a voltaic pile of his own. In one of his first experiments with this new apparatus, he immersed the wires leading from it in water; he discovered that, whenever the current was flowing, bubbles of gas were given off. They were bubbles of two gases, hydrogen and oxygen, and Nicholson realized that he had, in effect, reversed the process demonstrated by Cavendish 17 years earlier, in which he had produced water by burning hydrogen in the presence of oxygen. He had, in modern language, 'electrolysed' water; it was the first ever demonstration that an electric current could bring about a chemical reaction.

Nicholson was the editor of a chemical journal, and he lost no time in publishing an account of his discovery, which was thereby made known to the world before Volta had even got round to announcing his own invention. Nicholson's demonstration of the possibility of bringing about chemical reactions by means of an electric current, combined with Volta's demonstration of the generation of an electric current by chemical means, marked the birth of *electrochemistry.*

HUMPHRY DAVY In the early nineteenth century, Albemarle Street, just off London's Piccadilly, became the first in the city to be made a one-way street. The action was in part prompted by the jam of horse-drawn carriages that occurred whenever the Royal Institution staged one of its scientific lectures. The Royal Institution – a private, non-profit-making body – was created in 1800 by a wealthy amateur scientist and part-time spy named Benjamin Thompson, Count Rumford. Its object was to provide facilities for scientific education and research, and for the dissemination of scientific knowledge. Rumford equipped the Royal Institution with handsome premises in Albemarle Street, including a lecture theatre of great elegance, which serves the same function today. Soon afterwards he took off for Paris, where he became the lover of Lavoisier's widow. But before he left, he took the inspired decision to invite a young Cornishman named Humphry Davy, to become the Institution's assistant lecturer.

Davy was born in Penzance, Cornwall, in 1778. He was the son of a woodcarver, and was essentially self-educated. When he was 19, he read Lavoisier's *Elementary Treatise,* and this led to a lifelong love affair with chemistry. When Rumford invited Davy to lecture at the Royal Institution, he was working as medical superintendent of a fashionable spa in Bristol. Within a year, the professor of chemistry died, and Davy was appointed in his place. He was at this time a handsome, curly-haired, 23-year-old, and he was already a superbly gifted lecturer. It was the excitement aroused by his lectures that caused the traffic problems in the street outside. The less-than-wholly-scientific appeal of his public performances is conveyed by the comment of one well-born lady that, 'Those eyes were made for more than poring over crucibles.'

When he learned of the work of Volta and Nicholson, Davy became passionately interested in the subject of electrolysis. Many scientists at this time suspected that common substances such as magnesia, potash, and soda contained metallic elements yet to be identified. Davy constructed a powerful battery, with over 250 metal plates, and ran currents through solutions of all of them. From potash he obtained a previously unknown metal, which burst into flame on contact with water. He named this potassium. Just one week later, he obtained another from soda, which he named sodium. In the following year, he succeeded in isolating four more new elements: barium, strontium, calcium, and magnesium.

Davy's achievement in isolating potassium sent London society into a frenzy of hero-worship. The enthusiasm for his lectures was such that at one point tickets for his lectures were changing hands for £20 – more than £1,000 in today's money. In 1815, he crowned his career with the invention of the Davy lamp, which made it possible for miners to work safely in the presence of the deadly gas, firedamp. He ended his days rich and famous, president of the Royal Society, and a national treasure, with only one thing to mar his enjoyment: his jealousy of Michael Faraday, who was his own greatest discovery, and his successor at the Royal Institution.

HOW OLD IS THE EARTH? In the early nineteenth century, geology was at the cutting edge of science. As understanding of the succession and composition of the rocks increased, and as more and more fossil remains were uncovered, the question of the true age of the Earth became a matter of fierce debate. The age of 6,000 years implied by the study of biblical texts began to look more and more implausible, and scientists found themselves having to consider the possibility that the Earth's history went back a lot further than had previously been supposed. But one of the most startling estimates was put forward by a man who was not a geologist at all, but a mathematician.

Jean-Baptiste Joseph Fourier was born in Auxerre, France, in 1768. He was the son of a tailor, but was an orphan by the time he was eight. He played a minor part in the French Revolution, and narrowly escaped the guillotine. He formed an ambition to be an artillery officer, but was unable to achieve this ambition on account of his humble birth. However, thanks to the influence of his bishop, he was able to attend the military academy in Auxerre; on graduation, he was offered a position on the teaching staff. When the École Normale was founded in Paris in 1895, he became a lecturer; and his success in that post resulted in his being appointed professor of analysis at the École Polytechnique.

In 1798, he accompanied Napoleon to Egypt, and was made governor of a portion of the country. When Napoleon fell, Fourier not only survived, but received fresh honours at the hands of the restored Bourbon dynasty. In 1822, he was appointed joint secretary – with the anatomist Cuvier – of the Academy of Sciences.

Fourier's particular interest was in the way heat flowed from one body to another. This was a complex, and little understood, subject, involving not only the difference in temperature between two bodies, but in their shape and composition. Fourier brought his exceptional mathematical abilities to bear on the problem, and in 1807 he published what is now known as Fourier's theorem. It brought him instant fame, and it was in recognition of this that Napoleon made him a baron in 1808.

What Fourier showed was, that a complex periodical fluctuation, in which a situation regularly returned to its previous condition, could be broken down into a series of superimposed, simple fluctuations, which could be recombined to give the originally observed periodicity; and that the process could be summed up in the form of a mathematical series.

In 1822, he published his *Analytical Theory of Heat,* which still stands as one of the masterpieces of nineteenth-century science. In it, he applied his mathematical discoveries to the subject that had always fascinated him: the transfer of heat from one body to another. One problem to which he turned his attention was the question of how long it might have taken the Earth to cool to its present temperature. His calculations implied an age for the Earth of the order of 100 million years. This is only about $\frac{1}{50}$th of the figure accepted today. But it was a much greater number than anything envisaged by his contemporaries, some of whom were still trying to come to terms with the figure of 75,000 years that had been daringly advanced by the French naturalist Buffon 80 years earlier.

Fourier's theorem laid the foundations for a new branch of mathematics called *harmonic analysis*, which has applications in a wide range of situations where complex processes need to be broken down into their component parts so that they can be modelled and their outcome projected into the future. It is used in the analysis of sound waves and musical harmonics, in the study of variable stars and competition between animal species, and in the study of long-term climate change: anywhere, in fact, where phenomena of a periodic, wave-like character present themselves.

FOURIER AND THE GREENHOUSE EFFECT The term 'greenhouse effect' has a modern ring; but it is in fact nearly 200 years old. It was coined by Fourier.

After publishing *The Analytical Theory of Heat,* with its study of the cooling processes in the body of the Earth, Fourier turned his attention to the gases in the atmosphere, and the transfers of heat taking place within and through them. He provided an explanation of why the variations between daytime and night-time temperatures were so small; and he considered the processes that might have given rise to a climate in which animal, including human, life could have developed.

What Fourier proposed was that the Earth's atmosphere acted as an insulating blanket, slowing down the rate at which heat was radiated into space during the night, and therefore reducing the difference between daytime and night-time temperatures. It also had a similar effect in reducing the range of summer and winter temperatures. This was what he called the 'greenhouse effect'.

Fourier also created a personal 'greenhouse effect' of his own. He developed a morbid preoccupation with the maintenance of his body temperature. He kept his house seriously overheated, and went about swathed in layer upon layer of insulating clothing. Whether it would have contributed to the prolongation of his life will never be known, as he died from a fall down stairs when he was aged 62.

BERZELIUS AND CHEMICAL SYMBOLS The advice, 'Always quit when you're ahead', applies as much to science as to any other area of human activity. And there have been some eminent scientists who would have been more favourably remembered had they acted upon it. Young men who advance the boundaries of their disciplines sometimes become obstacles to progress in their old age. No doubt the same would be true of women, if they were ever to get the chance to run the show.

In the 1830s and 1840s, the Swede Jöns Jacob Berzelius bestrode the world of chemistry like a colossus, to use Shakespeare's phrase. His reputation was so great that whatever he said went. If he was wrong, as he often was, that was another setback for science. When he died in 1848, a weight was lifted off the shoulders of his younger contemporaries. Yet this same Berzelius had once been an innovator. He was one of the people who laid the foundations of modern chemistry, and in his early thirties he gave chemists a language they have been using ever since.

He was born in Väversunda Sörgård, near Linköping in southern Sweden, in 1779. His father, who was a clergyman, died when Berzelius was four, and his mother died when he was eight; but he had the good fortune to have another clergyman for his stepfather, who encouraged him to further his education. His stepfather did not, however, spoil him in the matter of money, and his years studying chemistry and medicine at Uppsala University were years of some hardship, which he softened by tutoring private students. In 1802, he graduated M.D., and was appointed assistant professor of botany and pharmacy in Stockholm. He later became professor of chemistry in the city's Caroline Institute of Chemistry and Surgery.

Berzelius was an adherent of Dalton's atomic theory. In 1807, when he was 28, he embarked upon a massive programme of investigation aimed at exploring the composition of chemical compounds, and establishing the atomic weights of the elements of which they were composed. During the next 10 years he analysed more than 2,000 compounds; and in 1818 he published a remarkably accurate table of atomic weights, showing the proportions in which the elements combined, which he improved on in a second version published in 1826. Even this table was marred by a residual confusion, however; Berzelius had never been able to sort out properly in his mind the difference between atoms and molecules.

At the time he was carrying on these researches, chemists were severely handicapped by their cumbersome symbolic language. It was a pictorial language derived from alchemy, which obscured, rather than reflected, the reality of chemical reactions. Dalton had tried to improve on it; but his symbols too were pictorial in nature, and very cumbersome. Without an efficient system of chemical symbols, chemists were in much the same position as mathematicians had been when they were restricted to using Roman numerals.

Berzelius cut through the tangle, and gave chemistry the symbolic language that, with minor changes, it has used ever since. And he earned the undying gratitude of textbook typesetters. His system had two features that streamlined the description of reactions, and greatly assisted thinking about the underlying reality. Firstly, he abandoned the use of the full names of the elements, substituting instead the first letter, or two early letters, of an element's Latin name. Copper (*cuprum*) became Cu, and gold (*aurum*) became Au. Secondly, he had the bright idea of describing compounds by combining the symbols for the individual elements of which they were composed. Zinc sulphide became plain ZnS. It was a simple system, and now seems obvious. But like that other symbolic system of great power, Arabic numerals, it wasn't obvious at all until it was invented.

Berzelius' later years were overshadowed by illness, but he was lionized wherever he went. And in 1835, at the ripe age of 56, he gave up single life in favour of marriage to a good-looking young woman of 24, the daughter of an old friend, who gave him 10 years of great happiness.

CHEMICAL FORMULAE An aspect of chemical notation that gave Berzelius some trouble was how to indicate the difference between compounds of the same elements. A carbon *monoxide* molecule is composed of an atom of carbon and an atom of oxygen, whereas a molecule of carbon *dioxide* is composed of one atom of carbon and two of oxygen. In his first scheme he used little dots to indicate the difference. He next tried algebraic symbols. His last version employed *superscripts*: little numbers above the symbol for the elements. It was two German chemists, Liebig and Poggendorff, who later changed these to *subscripts,* giving us the notation we use today; so that carbon monoxide is written CO, and carbon dioxide CO_2 .

FRAUNHOFER AND HIS LINES At the beginning of the nineteenth century, the stars were still objects of mystery. Their distance had been measured, their numbers had been counted, and their masses and motions had been computed. But they remained points of light, whose real nature was unknown. The telescope had added greatly to humanity's knowledge of the heavens. But knowledge is not the same as understanding. So long as astronomers had only the telescope to help them, they were like strangers in a great city, who knew the number, names, and appearances of their neighbours, but could not speak their language. Without new technology, nineteenth-century astronomy would have been more of the same: more stars, greater distances, but no greater insight. But as the century dawned, a new tool was developed, and with its aid astronomy was transformed. The tool was the spectroscope, and its inventor was a German optician named Joseph von Fraunhofer.

Fraunhofer was born in Straubing, Bavaria, in 1787. His father was a glazier, and Fraunhofer, the youngest of 11 children, was orphaned at 11 years of age. When he was 14, he was the only survivor when the tenement in which he lived collapsed. The elector of Bavaria, Maximilian I, hearing of his plight, gave him 18 ducats, which he used to set himself up as an optician. Self-taught, he made a study of the optical qualities of various kinds of glass, and became a master of the craft of optical instrument manufacture, and manager of the Munich Philosophical Instrument Company.

The study of the spectrum of visible light had marked time since Newton had broken it into its constituent colours over a century earlier. The lenses and prisms that Fraunhofer manufactured were superior to those that Newton had been able to produce; and in 1814 he used a telescope in conjunction with a prism in a precision instrument of his own invention: the *prism spectrometer*. With its aid, between 1814 and 1817, he made a number of discoveries that had eluded Newton. In particular, he discovered that the spectrum contained not just bands of various colours, but a number of fine dark lines. The bands of colour represented wavelengths of light that were present in the spectrum. The dark lines represented wavelengths that were absent. He counted 600 of these (now known as Fraunhofer lines), and included them in a chart, in which the most prominent, starting from the red end of the spectrum, were labelled A, B, C, etc.; the system is still in use today. He discovered that the lines in reflected sunlight from the Moon and planets occupied the same position as they did in direct light from the Sun itself. He even succeeded in analysing the spectrum of the light from a star, and observed that certain lines present in the Sun's spectrum were absent from the spectrum of the star. Had he been a recognized scientist, his discoveries would have created a stir. But he was only a craftsman, and academic snobbery prevented him from presenting his results to a learned society. It would be another 50 years before his work would be taken up by 'gentlemen'.

KIRCHHOFF'S EXPERIMENTS   Fraunhofer's work was to have enormous consequences for the study of astronomy, and it also prepared the way for later spectroscopic analysis of matter at the atomic level. But the essential work in turning his discoveries into a new science was done by a German physicist named Gustav Kirchhoff.

Kirchhoff, the son of a lawyer, was born in the Prussian city of Königsberg (now Kaliningrad, Russia) in 1824, and attended the university there. He spent 20 years as professor of physics at Heidelberg, and 12 years in a similar post in Berlin. During his career he made many important contributions to the development of mathematical and experimental physics, in particular in relation to the theory of electrical conductivity. He was the first person to demonstrate that electric current travelled with the velocity of light.

When Kirchhoff moved to Heidelberg in 1854, he made the acquaintance of a professor of chemistry named Robert Bunsen, 13 years his senior. Bunsen already had a distinguished career behind him – and the scars to prove it. His early research had been in organic chemistry. During one experiment, an explosion had cost him an eye; and he had twice almost died from arsenic poisoning. He later understandably abandoned organic chemistry, and refused to allow it to be studied in his department. His interests in inorganic chemistry were wide-ranging, and he was responsible for many inventions, of which the Bunsen burner is the best known.

In the 50 years since Fraunhofer had charted his lines, it had become clear that certain lines were associated with particular elements. If an element was present in a light source, it showed up as a series of bright lines – *emission lines* – in specific parts of the spectrum. If the element were not present, its absence was indicated by dark lines – *absorption* lines. Physicists now had a tool for analysing incandescent material, both in the laboratory and in distant, otherwise inaccessible sources.

Using Bunsen's new burner, which had the advantage of producing very little light of its own, Kirchhoff heated a number of materials to incandescence, and succeeded in 'labelling' the lines in their spectra with the names of the elements with which they were associated, producing a kind of spectroscopic bar code. Having determined the lines associated with the known elements, he and Bunsen were able to identify other lines as indicators of the presence of hitherto unknown elements. On 10 May 1860, they announced the discovery of a new element, caesium (from the Latin for 'sky blue'), and in the following year they discovered a second element, which they christened rubidium (Latin: red).

Kirchhoff's most important discovery was that, if light were passed through a gas, the resulting spectrum would display absorption lines in the same position as the emission lines the gas displayed when incandescent. This enabled him to deduce the presence of sodium in the atmosphere of the Sun, from absorption lines in its spectrum. Using this approach, other investigators were able to establish the presence of other terrestrial elements in the Sun's atmosphere.

One person who was not impressed with Kirchhoff's achievement was his banker, who asked, 'What use would it be to know there is gold in the Sun, if you cannot bring it down to Earth?' Some time later, Kirchhoff had the sweet revenge of going into the bank with a bag of gold sovereigns – a prize awarded by the British government – which he handed over with the words, 'Here is gold from the Sun.'

THE DISCOVERY OF HELIUM   One of the first triumphs of extraterrestrial spectroscopy occurred in 1868, when an element was discovered in the atmosphere of the Sun that was unknown on Earth. The person responsible was an astrophysicist named Norman Lockyer, who was at the time on the staff of the British War Office. Lockyer was engaged on a study of the solar prominences (the enormous flares that shoot out from the surface of the Sun). In October 1868, he observed a yellow line in the solar spectrum that was separate from the lines he had previously identified with sodium, and which did not match the spectrum of any known element. He concluded that the line was evidence of an as-yet-undiscovered element, which he named helium, from the Greek word for the Sun. The suggestion was dismissed by other chemists but, 27 years later, in 1895, William Ramsay, professor of chemistry at University College, London, succeeded in isolating Lockyer's gas from a specimen of a radioactive mineral called cleveite. Lockyer lived long enough to see his proposition vindicated.

Ramsay was not, in fact, the first person to observe helium in the laboratory. An American geologist named W.F. Hillebrand had obtained a similar result four years before him, but had put it down to a defect in his equipment.

MICHAEL FARADAY In 1812, some 10 years after Humphry Davy had begun lecturing at the Royal Institution, a 20-year-old apprentice in a local bookbinder's shop was given tickets for four of his lectures. His name was Michael Faraday. He was one of ten children of a blacksmith, and he had been born in a village called Newington, on the outskirts of London. He had had little formal schooling but, from the age of 13, when he began his apprenticeship – and with the encouragement of a kindly employer – he read widely; he managed to acquire a good knowledge of science from the books that passed through his hands. The tickets for Davy's lectures were a gift from the father of a friend, who had been impressed with the notes that Faraday had made on his scientific reading.

These lectures changed the course of Faraday's life. A few months later, Davy was temporarily blinded by an experiment that went wrong, and Faraday contrived to be taken on as his temporary secretary, while continuing with his own work. When this employment ended, he wrote to Davy asking for a full-time job, and supporting his application with the bound notes from the lectures he had attended earlier in the year. By a stroke of luck, Davy's assistant was shortly afterwards dismissed for brawling, and Faraday was taken on in his place.

It was a lowly post, involving much fetching and carrying, but he worked hard and was soon performing the most intricate chemical experiments. Soon after their collaboration began, Davy became Sir Humphrey, married a rich widow, and resigned his lectureship. Faraday accompanied them on an extended visit to the Continent in the capacity of scientific assistant and general dogsbody. The trip involved him in much humiliation at the hands of Davy's snobbish young bride, but it gave him the equivalent of a university education. He learned French and Italian, and made the acquaintance of many men of science. These included, in France, the chemist Gay-Lussac and the physicist Ampère, and in Italy the 70-year-old physicist Alesssandro Volta, who was then at the height of his fame.

FARADAY AND ELECTROLYSIS Compared with many great men of science, whose most important work was done in their twenties, Faraday was slow to get into his stride. He is recognized now as one of the most important scientists of the nineteenth century and, as a physicist, as one of the greatest of all time. But had he died in his early thirties, his name would be virtually unknown. He came late to education, and mathematics, that quintessential young person's science, would always be a closed book to him. He was compensated for his lack of mathematical ability by a phenomenal gift for visualizing natural processes, as if they were physical events taking place before his eyes; it was this gift that would lead to many of his most important insights.

Within months of his return to the Royal Institution, he achieved his first notable success, when he developed a method by which gases such as chlorine and carbon dioxide could be liquefied under pressure. Two years later he made a major contribution to the development of organic chemistry, when he discovered benzene, a compound that would later play a key role in the explanation of molecular structure.

In 1825, he was made director of the laboratory at the Royal Institution; he began a programme of research into electrochemistry – the subject Davy had pioneered. Davy had isolated a number of metals by passing an electric current through compounds containing them. Faraday named this process *electrolysis,* and the metal rods inserted in the solutions *electrodes*. In 1832, he formulated his *laws of electrolysis,* which established for the first time the relationship between electricity and chemical processes, and expressed that relationship in precise quantitative terms.

ELECTRICITY AND MAGNETISM   While pursuing research into the relationship between electricity and chemistry, Faraday simultaneously addressed himself to the relationship between electricity and magnetism. His interest had been aroused, when he was a teenager, by an article in *Encyclopedia Britannica*, which inspired him to construct a number of electrical devices. One of these was a voltaic pile, the battery described by Volta in 1800.

In the early years of the nineteenth century, many scientists suspected that there might be a connection between the newly discovered phenomenon of electricity and the more familiar phenomenon of magnetism. In 1820, a 42-year-old Danish physicist named Hans Christian Oersted set the debate alight with his description of a classroom experiment in which he had brought a compass needle close to a wire through which an electric current was passing.

Oersted was professor of physics and chemistry at the University of Copenhagen. He was a scientist with an unusually wide range of interests, and a believer in the fundamental unity of the laws of nature. Encouraged by this belief, he made many efforts to establish connections between chemical and magnetic forces and light, and to show that they were all essentially electrical in character.

In the experiment in question, Oersted discovered that a compass needle was deflected at right angles to the direction of the passing electric current, and that if the current were reversed, the needle was deflected by the same amount in the opposite direction. The publication of this result created a sensation. On hearing of Oersted's experiment, Faraday embarked on a programme of follow-up investigations aimed at elucidating the nature of the relationship between electricity and magnetism. He devised an experiment in which a movable wire was wound round a fixed magnet, and a movable magnet was able to pivot around a fixed wire. When a current was passed through the apparatus, the movable magnet began to pivot about the fixed wire, and the movable wire began to pivot about the fixed magnet.

Faraday next attempted to reverse Oersted's experiment. Oersted had, in effect, used an electric current to create magnetic attraction. Faraday now set out to establish whether it was possible to use a magnetic field to generate an electric current. He succeeded, and in doing so, he created the world's first transformer.

THE CHRISTMAS LECTURES Though childless, Faraday and his wife loved children, and in 1826 he advertised a series of six talks for children over the Christmas period. These were a huge success, beginning a tradition that continues to the present day. Then, as now, these talks presented the latest scientific ideas in terms that took the intelligence of the audience for granted, enabling them to enjoy the thrill of scientific discovery in the company of the finest scientists of the day. One of these talks, *The History of a Candle,* was so successful that Faraday was forced to repeat it at regular intervals, and it became famous as a classic of popular science.

In the year after these talks were introduced, he was offered the position of professor of chemistry at London's newly established University College, but his loyalty to the Royal Institution, to which he owed so much, led him to decline it. He was offered a knighthood, and declined that too. Principled to the end, he remained plain Mr Faraday. And when he died, in 1867, he was buried in a simple grave in Highgate Cemetery.

FIELDS OF FORCE Faraday's experiments with electricity and magnetism would later lead to the development of the electric generator and the electric motor. In a very real sense they represented the moment of conception of our modern world. But it is a mistake to judge the significance of his research work solely by the technology to which it ultimately led, or to exaggerate his own contribution to the development of that technology. The road – from those experiments in the basement of the Royal Institution to today's power stations and electric trains – was a long and twisting one; the journey was only made possible by the efforts of a host of gifted scientists and engineers. Faraday's crucial contribution to the study of electromagnetism was not the toys he constructed, but the *understanding* he arrived at; it was in this that his gift for visualization achieved its greatest triumph.

It was not enough to have *demonstrated* the relationship between electricity and magnetism. He wanted to understand the forces behind them. The French physicist André Marie Ampère had already demonstrated that a magnetic force emanated from a wire through which an electric current was passing. Faraday discovered that if iron filings were scattered on a sheet of paper under which a magnet had been placed, and the paper gently tapped, the filings would assemble themselves in a pattern of lines radiating out from the magnet. To Faraday's intensely visual imagination, these lines suggested more than a local disturbance within, or immediately surrounding, the magnet. They spoke of *fields of force,* spreading in every direction, diffusing in accordance with Newton's inverse-square rule, and continuing to the uttermost limits of the universe. It was an image that made no great impression on his immediate contemporaries. But a younger man – the physicist James Clerk Maxwell – could express Faraday's vision in the form of mathematical laws. In his hands it was destined to lead to the greatest revolution in physics since Newton's *Principia*.

FARADAY THE SHOWMAN Faraday was a modest man, who had little interest in public acclaim. Yet, paradoxically, his gift of visualization, and his desire to share the fascination of science, made him as successful a public lecturer as Davy had been. And, in his quiet way, he was a consummate showman.

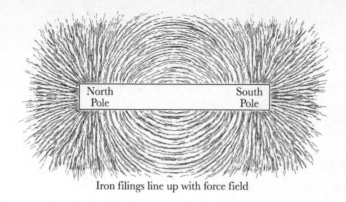

North Pole   South Pole

Iron filings line up with force field

Figure 12. The Fields of Force Surrounding a Magnet, As Demonstrated by Faraday.

He was, without doubt, one of the most gifted popularizers of science who has ever lived. Charles Darwin, who was made ill by the slightest excitement, and avoided public events like the plague, was a regular attendee at his lectures. Charles Dickens, whose own public performances were a byword for drama and suspense, was a devoted fan.

The showmanship that captivated Faraday's audiences is well conveyed by the way he illustrated a lecture on electrical conductivity. He had a wooden cage made in the shape of a 3.6-metre/12-foot cube, and covered it with metal foil and metal wire. When the moment came to make his point, he stepped quietly into the cage with his measuring instruments, and instructed his assistant to charge it up to around 100,000 volts. Sparks flew in every direction, and the audience united in a sharp intake of breath, while Faraday calmly carried on with his presentation.

OWEN AND THE DINOSAURS A younger contemporary of Faraday's was the zoologist Richard Owen, who for a period in the middle of the nineteenth century enjoyed a towering reputation in British science. One of his many achievements was the creation of London's magnificent Natural History Museum. Unfortunately, he made the mistake of backing the wrong side in the debates between Darwin's supporters and their opponents, and his reputation never recovered from the drubbing he endured at the hands of a younger generation of scientists. To these men he represented the 'old brigade', the men they considered were holding back scientific progress.

Owen is remembered as the man who coined the word *dinosaur* to describe a new class of animal. He had identified it on the basis of the fossilized remains of three very different creatures, all discovered in the limestone deposits of southern England. The name 'dinosaur' was derived from two Greek words meaning 'terrible lizard'. 'Terrible', a powerful word in 1842, meant 'inspiring terror'; by choosing this name, Owen wanted to emphasize the perfection of these amazing creatures. A passionate opponent of evolutionary ideas, it was his way of refuting those who saw evolution as a process of advance from 'lower' to 'higher' forms of life. He would have been dismayed had he foreseen that his magnificent monsters would end up as cuddly toys and cartoon characters.

Owen's suggestion that three such diverse creatures were related, and that they represented a group hitherto unknown to science, was a daring flight of the imagination. But subsequent discoveries amply vindicated his judgement; and he deserves an honoured place in biology's Hall of Fame.

THE EXTINCTION OF THE DINOSAURS   For over 100 million years, from about 180 million to 65 million years ago, the dinosaurs were a dominant form of life on land. Living when the present-day continents were joined in one vast super-continent, their remains are found all around the world. Most of them were large, and some were very large indeed: the largest land animals that have ever existed. Not all of them were meat-eaters. Many of the biggest were herbivores. And then they 'vanished' from the face of the Earth. 'Vanished', that is, in a geological sense. They did not, literally, die out overnight. But in terms of the timescale by which geological history is measured, the end came very quickly indeed. It occurred at the end of the Cretaceous period, and the beginning of the Tertiary: what geologists call the K/T boundary (*Kreide* is German for Cretaceous).

Until the 1980s, this sudden disappearance was one of the great mysteries of science. In the early twentieth century, only the Germans paid much attention to it. In the 1950s, scientists elsewhere started to take it seriously. In the absence of hard evidence, some wacky theories were bandied around. Some people suggested that the dinosaurs had got too big for their boots or, rather, for their little brains. Others invoked a sudden climatic shock, involving a sharp fall in global temperatures. Some even suggested that the dinosaurs had died out because little mammals ate their eggs.

In 1977, scientists from the University of California at Berkeley, led by the geologist Walter Alvarez and his physicist father Louis Alvarez, were examining rocks near the town of Gubbio, central Italy. Precisely at the K/T boundary, they discovered massive quantities of the element iridium. Iridium, although rare on Earth, is frequently found in meteorites. This led the Alvarezes to suggest that it was the remnant of an asteroid impact. They further speculated that, if the amount found at Gubbio was an indication of the amount likely to be found in other localities, it implied an impact at some unknown location by a meteorite at least 10 kilometres/6 miles in diameter. An impact on this scale would, they said, have raised clouds of dust that would have shut out the Sun's heat for a year or more, creating a worldwide 'nuclear winter', with devastating implications for the survival prospects of many forms of life. In the years that followed, similar iridium signatures were found in K/T boundary rocks in more than 150 locations around the globe.

But, if there had been such an event, where was the crater? It might have been under the ocean, in which case it was unlikely ever to be found. In 1981, two geologists prospecting for oil had discovered a massive crater at Chicxulub on the coast of the Yucatán peninsula in Mexico. It was of no professional interest to them, and its existence was not publicized at the time. But as the search for a K/T impact site intensified, evidence began to accumulate that the Chicxulub crater was indeed the remnant of an impact of exactly the scale and date suggested by the Alvarezes.

It is now accepted that the Chicxulub impact was a major, if not the only, factor in the extinction of so many life forms, on land and in the sea, that occurred 65 million years ago. It was not just the dinosaurs that disappeared. Many groups of animals and plants that had flourished for millions of years vanished with them. And their disappearance made possible a diversification of other life forms, including the mammals and flowering plants that now abound in the lands the dinosaurs once ruled.

The K/T event was the most recent of four great extinctions during the past 500 million years. It is unlikely to be the last.

THE HEAVIEST ANIMALS – ON LAND AND IN THE SEA

The claim of the larger species of dinosaur to the title of the biggest creatures ever to walk the Earth is beyond dispute, as a comparison of some of them with the land animals alive today makes clear. Diplodocus measured 30 metres/90 feet from head to tail. Brontosaurus weighed in at 35 tons. Brachiosaurus stood 10-metres/30-feet high. The heaviest recorded land animal of modern times – an African elephant killed in Angola in 1955 – was a mere 4 metres/13 feet high at the shoulder, and tipped the scales at an anorexic 12 tons. But animals that live in water can grow to a size that makes even Brontosaurus seem undernourished. There are blue whales swimming the seas today that weigh more than 150 tons.

JAMES CLERK MAXWELL

The names of Einstein, Darwin, and Newton are known throughout the world, but there are some whose names are virtually unknown to the general public, even though their achievements were almost in the same league. This is the case with the Scottish physicist James Clerk Maxwell. Professional scientists, and physicists in particular, recognize him as one of the most brilliant and influential scientists who have ever lived, but outside scientific circles his name is not widely known.

He was born in Edinburgh in 1831. Despite the death of his mother when he was 8, he had a happy childhood. He showed exceptional promise from an early age, especially in mathematics. When he was 15, he submitted a mathematical paper to the Royal Society of Edinburgh, the standard of which astounded those who read it. In the following year he had the good fortune to make the acquaintance of the 70-year-old physicist William Nicol, who also lived in Edinburgh. Nicol had done important work on the use of crystals to investigate the nature and behaviour of light, and Maxwell's teenage conversations with him led to a lifelong interest in the subject of light, and of other forms of radiation.

He studied mathematics with distinction at Cambridge, and it was while he was a student there that he had the defining intellectual experience of his life: reading Faraday's *Experimental Researches in Electricity*. While he was still a student he made a major contribution to the development of the subject with a brilliant published paper entitled *On Faraday's Lines of Force*. In 1856, at the age of 25, he was appointed a professor at Marischal College in Aberdeen; and in 1860 he moved to Kings College, London, as professor of natural philosophy and astronomy. It was around the time of his move to London that he made his first major contribution to the advancement of physics.

HEAT IS MOVEMENT The problem that Maxwell addressed himself to in 1860 was the behaviour of gases, especially in relation to changes in temperature. There was at this time still no real understanding of what heat was; the idea that it was some kind of fluid that passed from warm bodies to cool still had considerable support. Maxwell addressed it as a question of the behaviour of molecules in rapid motion; his outstanding mathematical skills enabled him to treat the problem statistically. He developed an equation that described the distribution of velocities among the molecules of a gas at a given temperature. This equation showed that, although some molecules moved comparatively slowly, and some exceptionally fast, the majority moved at a middling sort of speed that increased as temperature rose, and decreased as temperature fell. Temperature, and heat itself, were produced by the motions of molecules, and this was as true of liquids and solids as it was of gases. Heat was not something that flowed from one place to another; it was simply another word for the activity of the molecules of the substance in question.

MAXWELL'S EQUATIONS In 1871, Maxwell reluctantly accepted the newly created position of professor of experimental physics at Cambridge. It was not a post for which he was well fitted. He was a theorist, rather than an experimenter; his highly mathematical lectures were a turn-off for all but a tiny handful of students who were capable of following his reasoning. But it was while occupying this chair that he produced the work for which he is most famous; work that provided an inspiration for a generation of scientists such as Einstein who explored the implications of his ideas.

Faraday's imagination had pictured electricity and magnetism as invisible fields of force extending out into space. But it was Maxwell who gave the idea the mathematical form that made it usable. He proved that electricity and magnetism were intimately related, and that the one could not exist in the absence of the other. In 1864, he published a paper entitled *A*

Dynamical Theory of the Electromagnetic Field, in which he set out four simple equations – still known as *Maxwell's equations* – that explained almost everything about the relationship between them. The work that Maxwell did between 1864 and 1873 laid a substantial part of the foundations of modern physics. Among other things, he showed that electric fields and magnetic fields move forward together, as *electromagnetic waves*. He established that these waves travelled at the speed of light, and he suggested that light itself was just one variety of a whole spectrum of electromagnetic radiation (*see figure 9, page 74*), that probably included other kinds of radiation not yet discovered. In 1873, he crowned his career with the publication of his *Treatise on Electricity and Magnetism,* one of the greatest monuments any man of science ever raised to his own genius.

LONG-WAVE RADIATION Maxwell died in 1879, not yet 50, and too soon to see his forecasts vindicated. Had he lived for just another 10 years he would have known of the discovery of long-wave radiation with a wavelength 1 million times greater than that of visible light, the radiation we know as radio waves. It was discovered by the German physicist Heinrich Hertz, who was able to show that it travelled at the speed of light, and was reflected and refracted like light.

Our modern world is filled with a vast range of electromagnetic radiation, as Maxwell suspected: infrared rays, X-rays, radio waves, gamma rays, microwaves. And we handle them all with the equations he gave us.

ORDERING THE ELEMENTS Once in a while, a scientist comes along who suggests a new way of thinking. When this happens, we talk about science having been provided with a new *paradigm*: a new model of the natural world. The paradigm that made sense of chemistry, and still supplies the framework of the science, is the periodic table, which has its origins in the work of the Russian chemist Dmitri Mendeleyev.

Mendeleyev was born in Tobolsk, in western Siberia, in 1834, the youngest of 14 children. His father was headmaster of the local high school, but he became blind in the year Mendeleyev was born. His mother was the daughter of a factory-owner, and she reopened one of her father's factories to help support her family. The young Dmitri had no interest in school, but he was inspired with a love of science by a private tutor.

When he was 13, his father died, and his mother's factory burned down. Having no reason to remain in Siberia, and wishing to further her son's education, she set off on the 2,000-kilometre/1,300 -mile journey to Moscow, with Dmitri and an older daughter. In Moscow, he was refused entry to the university; so they travelled a further 650 kilometres/400 miles to St Petersburg, where a friend of his father obtained a scholarship for him to study science at the Central Pedagogical Institute, attached to the university. Within a year, his mother and his sister died, and he was admitted to the institute hospital, suffering from tuberculosis. He was given two years to live, but survived.

After a long stay in hospital, he qualified as a teacher, and became an unpaid lecturer at St Petersburg University, dependent upon fees from private students. When he was 22, he obtained a grant to study abroad. He went first to Paris; and then to Heidelberg, where he was lucky enough to meet

Bunsen and Kirchhoff, who were conducting the researches that would lay the foundations of spectroscopy.

In September 1860, he travelled to Karlsruhe, in Germany, to attend the first International Congress of Chemistry, which had been arranged to settle a dispute about how best to arrive at the weights of the individual elements. It was attended by 140 of the world's most eminent chemists; and the speeches he heard there created an interest that endured for the rest of his life.

FRANKLAND AND CANNIZZARO In 1860, chemistry was still in confusion. In the 50 years since Dalton had outlined his atomic theory, a number of chemists, most notably Berzelius, had built upon the foundations he had laid down. But there was still no general agreement on the most basic aspects of the science. The confusion was such that there were up to 20 different formulas in use to describe quite simple compounds.

A significant contribution to the regularization of the subject had been made by the English chemist Edward Frankland. Frankland, who was born in Lancashire in 1825, was a pharmacist's apprentice who had taught himself chemistry; to such good effect that he had subsequently obtained a Ph.D. from the university of Marburg in Germany, and become professor of chemistry at Owens College in Manchester. In 1852, he had introduced the concept of *valency*: the idea that the atoms of each individual element had their own specific capacity for combining with the atoms of other elements, and that this determined the ratios in which they joined together to form compounds. Thus, hydrogen has a valency of 1, and oxygen a valency of 2, so that one atom of oxygen will combine with two atoms of hydrogen to form a molecule of water, which is written H_2O. Similarly, a carbon atom, which has a valency of 4, will combine with two atoms of oxygen, with a valency of 2, to form one molecule of carbon dioxide, or CO_2. Valency subsequently became a useful tool in the day-to-day work of the chemist – but quite why the elements possessed this property would not become clear for five decades.

An important contribution to the understanding of the elements had been made by another speaker at the conference, the Italian Stanislao Cannizzaro. Cannizzaro, the son of a magistrate, was born in 1826, in Palermo, Sicily. He had had a colourful career, including exile in Paris for his part in an uprising against the King of Naples in 1848. But he was later able to return to Italy, and at the time of the conference he was a professor of chemistry in Genoa. In 1858, he had published a pamphlet in which he had established for the first time the crucial distinction between atoms and molecules.

MENDELEYEV'S TEXTBOOK Cannizzaro's speeches in Karlsruhe had a powerful effect on Mendeleyev; he returned to Russia convinced of the truth of Cannizzaro's assertion that the only rational measure of the weight of an element was the weight of its individual atoms. This belief would inspire his future researches.

On his return to St Petersburg in 1861, he obtained a teaching post in the Technical Institute, and quickly became an evangelist for the latest ideas in chemistry, which were virtually unknown in Russia. Discovering that there was no Russian textbook in organic chemistry (the chemistry of the compounds that form the basis of living matter), he proceeded to write one – in two months flat.

In 1866, when he was 32, he became professor of chemistry at the university. Soon afterwards he set about writing a text-book entitled *The Principles of Chemistry*; the first volume appeared in 1868. It was a book that would be translated into a score of languages, and become the standard text for two generations of students. It was while he was writing the second volume that he made the breakthrough that brought order to the elements, and ensured his undying fame.

MENDELEYEV'S DREAM It had been known for some time that some elements shared similar properties; and chemists had begun to wonder whether it might be possible to classify them, as Linnaeus had classified animals, on the basis of these. In 1864, an English chemist, John Newlands, had drawn attention to the fact that, if the elements were arranged in the order of their atomic weights, the resulting table displayed a *periodicity*, meaning that similar characteristics recurred at more or less regular intervals. He expressed this idea in a rule he called the *law of octaves*, since similar characteristics seemed to recur after every eight places in his table. But when he announced his 'discovery' at a meeting of chemists, he was ridiculed.

Mendeleyev was aware of Newlands' work, but was not impressed with the way in which it was expressed. In particular he disliked the way some elements seemed to have been shoe-horned into their places to preserve an impression of period-icity. As he began the second volume of his textbook, he tried to find an arrangement that would provide a framework for understanding the relationship of one element to another, but which would be free of the defects he perceived in Newlands' scheme. He was convinced that chemistry could not really claim to be a science until one could identify fundamental principles underlying its practice.

The organizing principle of his book was the grouping together of elements according to shared properties. By February 1869, he had written two chapters of his second volume, and was mulling over which group of elements he should write about next. He was under great pressure. His reflections on the ordering of the elements had given him the feeling that the principle he was seeking was almost within his grasp. He had written out the names and weights of the known elements on a series of cards, which he kept rearranging, as if he were playing patience. He was obliged to make a trip to the country; he was afraid that if he did not find the solution before he left, his concentration would be lost, and his chance of finding one would be lost with it. For three days, and most of three nights, he wrestled with the problem, until he was befuddled with lack of sleep. On the day he was supposed to depart, he fell asleep at his desk. While he slept, his brain continued shuffling the cards; when he woke, he realized he had the solution.

THE PERIODIC TABLE The secret that Mendeleyev's unconscious mind had glimpsed while he was asleep was that the elements could be arranged in horizontal rows in ascending order of their atomic weight, and in vertical columns according to their chemical characteristics – *leaving gaps where the pattern seemed to require them.*

He published these ideas in a paper entitled *On the Relation of the Properties to the Atomic Weights of the Elements.* This contained his periodic law, which stated that if the known elements were listed in order of ascending atomic weight:

1. They displayed a repeating pattern of rising and falling valency (the ratio in which they combined with other elements),

and

2. They formed groups that displayed a recurring pattern of other characteristics.

One consequence of Mendeleyev's discovery was that he was able to reposition 17 elements within his table on the basis of their chemical properties, implying that their accepted atomic weights were incorrect. He was also able, on the strength of gaps in his table, to postulate the existence of three elements then unknown, and even to forecast their properties.

The initial reaction to Mendeleyev's paper was as guarded as that afforded to previous attempts to bring order to the elements. But when it was discovered that the accepted atomic weights of some elements were indeed incorrect, his ideas began to be taken more seriously. And within 15 years, all three of the gaps in his table were filled: by the discovery of gallium (1875), scandium (1879), and germanium (1886), which each possessed the characteristics he had predicted. Although not the first to suggest that it might be possible to arrange the elements in an order that displayed periodicity, unlike his predecessors, Mendeleyev was able to show that there was an underlying logic that dictated such a table.

In 1876, at the age of 43, after many years in an unhappy marriage, he divorced his first wife. Under Russian law he was not permitted to remarry for 7 years. But he had fallen in love with a beautiful young art student of Cossack extraction. Unable to wait, he married her, and was charged with bigamy.

But the Tsar refused to punish him, saying, 'Mendeleyev has two wives, but Russia has only one Mendeleyev.' This second marriage was a happy one. It brought him two daughters and two sons, whom he loved; and years of productive work, in a study furnished with his wife's drawings of his heroes: Newton, Faraday, and Lavoisier.

THE PHYSICS BEHIND CHEMISTRY Mendeleyev's table has been modified since he constructed it. The modern version (*see page 170*) reflects the knowledge acquired since his day. It also contains 109 elements, compared to the 63 he knew of. But it is still recognizably his table, because he had hit upon the fundamental relationship between the elements, even though he had no idea how their atoms were put together.

Elements 1 (hydrogen) to 92 (uranium) are *natural* elements. These are the basic ingredients out of which the world is made. The rest are man-made. All the elements are built out of a few extremely small *elementary particles* called protons, neutrons, and electrons. Every atom of every element has a nucleus made up of protons and neutrons. Around this, the electrons spin, like planets around the Sun. Just as the Sun accounts for most of the mass of the solar system, so the nucleus accounts for most of the mass of the atom. And, just as the planets are separated from the Sun by vast empty spaces, so the orbits of the electrons are separated by vast empty spaces from the central nucleus. It is the number of neutrons and protons the nucleus contains that determines the atomic weight of an element (a proton weighs 1,836 times more than an electron). But it is the number and arrangement of the electrons that

The Periodic Table

This is the present-day version of the table that Mendeleyev constructed.

The numbers are the *atomic numbers* of the elements, and correspond to the number of protons in the nucleus.

Roughly speaking, the *atomic weight* increases as the atomic number increases. The exact weight depends on the number of protons *and* neutrons in the nucleus.

The vertical columns are called groups, and contain elements with similar chemical properties. These properties are determined by the number of electrons circling the nucleus, which is equal to the number of protons.

The names and symbols given for elements 104 to 109 are those currently approved by the International Union of Pure and Applied Chemistry (IUPAC).

1 H Hydrogen																	2 He Helium
3 Li Lithium	4 Be Beryllium											5 B Boron	6 C Carbon	7 N Nitrogen	8 O Oxygen	9 F Fluorine	10 Ne Neon
11 Na Sodium	12 Mg Magnesium											13 Al Aluminium	14 Si Silicon	15 P Phosphorus	16 S Sulphur	17 Cl Chlorine	18 Ar Argon
19 K Potassium	20 Ca Calcium	21 Sc Scandium	22 Ti Titanium	23 V Vanadium	24 Cr Chromium	25 Mn Manganese	26 Fe Iron	27 Co Cobalt	28 Ni Nickel	29 Cu Copper	30 Zn Zinc	31 Ga Gallium	32 Ge Germanium	33 As Arsenic	34 Se Selenium	35 Br Bromine	36 Kr Krypton
37 Rb Rubidium	38 Sr Strontium	39 Y Yttrium	40 Zr Zirconium	41 Nb Niobium	42 Mo Molybdenum	43 Tc Technetium	44 Ru Ruthenium	45 Rh Rhodium	46 Pd Palladium	47 Ag Silver	48 Cd Cadmium	49 In Indium	50 Sn Tin	51 Sb Antimony	52 Te Tellurium	53 I Iodine	54 Xe Xenon
55 Cs Caesium	56 Ba Barium	72 Hf Hafnium	73 Ta Tantalum	74 W Tungsten	75 Re Rhenium	76 Os Osmium	77 Ir Iridium	78 Pt Platinum	79 Au Gold	80 Hg Mercury	81 Tl Thallium	82 Pb Lead	83 Bi Bismuth	84 Po Polonium	85 At Astatine	86 Rn Radon
87 Fr Francium	88 Ra Radium	104 Rf Rutherfordium	105 Db Dubnium	106 Sg Seaborgium	107 Bh Bohrium	108 Hs Hassium	109 Mt Meitnerium										

57 La Lanthanum	58 Ce Cerium	59 Pr Praseodymium	60 Nd Neodymium	61 Pm Promethium	62 Sm Samarium	63 Eu Europium	64 Gd Gadolinium	65 Tb Terbium	66 Dy Dysprosium	67 Ho Holmium	68 Er Erbium	69 Tm Thulium	70 Yb Ytterbium	71 Lu Lutecium
89 Ac Actinium	90 Th Thorium	91 Pa Protactinium	92 U Uranium	93 Np Neptunium	94 Pu Plutonium	95 Am Americium	96 Cm Curium	97 Bk Berkelium	98 Cf Californium	99 Es Einsteinium	100 Fm Fermium	101 Md Mendelevium	102 No Nobelium	103 Lr Lawrencium

Figure 13.

determine an element's chemical properties, because when atoms combine, it is their electrons that do the joining.

The numbers in the periodic table are *atomic numbers*; they represent the number of protons in the nucleus. They also correspond to the number of electrons orbiting the nucleus, because every atom contains an equal number of protons and electrons. The electrons have a negative charge, which is balanced by the positive charge possessed by the protons. The *atomic weight* of an element depends on the total number of protons and neutrons in the nucleus. This tends to increase with increasing atomic number, but some elements come in multiple versions, called *isotopes*. For example, natural uranium (atomic number 92) comes in two forms: uranium 235, with 92 protons and 143 neutrons and an atomic weight of 235; and uranium 238, with 92 protons and 146 neutrons, and an atomic weight of 238 (equal to 238 hydrogen atoms).

The vertical columns are referred to as 'groups': these are families of elements with similar properties. Thus, the right-hand column contains the 'noble' or 'inert' gases: helium, neon, etc. These have been called the 'lazy' gases (*argos* is Greek for lazy), because they are slow to combine with other elements. This makes them useful for filling balloons (helium is safer than hydrogen), and fluorescent lamps (argon).

WHAT MAKES THE SKY BLUE? In 1854, when Faraday was still director of the Royal Institution, a young Irish-born physicist named John Tyndall was appointed a professor there. His special interest was in the behaviour of gases, and the way in which they conduct heat. He also made important discoveries concerning the behaviour of light as it passed through various substances. A by-product of his research was his explanation of the colour of the sky. It had already been established that liquid oxygen was blue; and this suggested that the colour of the sky might be attributable to the oxygen in the atmosphere. Tyndall, drawing on the work of the English physicist, Lord Rayleigh, was able to explain the blue colour as the result of the scattering of light by dust particles in the atmosphere. Rayleigh had shown that the degree to which light is scattered is in inverse proportion to the fourth power of the light's wavelength. This means that, for example, the violet element of sunlight, with a wavelength half that of red light, is scattered 2^4 times, that is, 16 times as much as the red element. It is this scattering of the blue/violet light that makes the sky blue. But Tyndall's explanation was modified by Albert Einstein 50 years later, when he proved that it was the molecules of the air, not particles of dust, that were scattering the light.

GLOBAL WARMING Tyndall's research into the transfer of heat by gases enabled him to add support to Fourier's theory of 'greenhouse effect'. In 1860, Tyndall measured the absorption of radiation by the atmosphere, and concluded that it was carbon dioxide and water vapour, rather than nitrogen and oxygen, that were responsible for the heat-retaining effect Fourier had described.

In 1896, the Swedish chemist Svante Auguste Arrhenius (*see page 141*) developed this idea further. He addressed the question of whether changes in ground-level temperatures might be related to changes in the levels of these heat-absorbing gases in the atmosphere. He calculated the effect that a doubling of the amount of carbon dioxide in the atmosphere might have on the Earth's climate. The answer was an increase in average global temperatures of between 5 and 6°C/41 and 43°F.

Arrhenius' thoughts on the subject of global warming caused no great stir at the time, because most scientists were not convinced that such an increase in the carbon dioxide content of the atmosphere was a plausible scenario. However, by the early 1940s, the British physicist G.S. Callendar had concluded, from a study of records from weather stations around the world, that a process of global warming, caused by increased carbon dioxide levels, was indeed occurring. Even then, many scientists found it difficult to believe that human activities could influence global temperatures in this way. It was not until the 1960s that opinion began to change, in the face of mounting evidence that both carbon dioxide levels and global temperatures had been rising for the best part of a century.

Average global temperatures are now known to have risen by over half a degree celcius – about one degree Fahrenheit – during the past 100 years, and analysis of past atmospheric conditions indicates that carbon dioxide levels in the Earth's atmosphere are the highest they have been for 750,000 years. Given that for most of the past 100 years, the rate of fossil-fuel burning has been only a small fraction of present levels, it is beginning to look as though Arhennius' estimate of the increase in global temperatures could be achieved before the end of the present century. The consequences of such an increase in average temperatures would be devastating.

DATING THE PAST Before the nineteenth century, the past was a matter of either 'history' or a vague 'prehistory' – things that had happened, or were believed to have happened, at some unspecified time before records were kept. History had sequence and depth. It was clear that the Fall of Rome had occurred before the Fire of London; one also knew how many years had elapsed between them. But prehistory had little sequence, and no depth. Everything before records were kept had simply happened 'a long time ago'.

We owe our present-day view of the past to two sciences: *stratigraphy* and *geochronology*. Stratigraphy is the study of the sequence of rocks or other deposits in the Earth's crust. Geochronology is the science of putting dates to them.

Stratigraphy came first. As archaeologists excavated the sites of ancient towns such as Troy and Jericho, they became skilled at identifying the sequence of successive settlements on the same site, and the specific character of the pottery, tools, etc., associated with each level. This enabled them to build up a picture of individual cultures, and the changes they underwent over time. Stratigraphy was also important in establishing the relative age of human remains predating civilization. Careful study of the deposits within which they were found made it possible to determine the sequence in which they had been laid down, and to correlate remains from widely separated locations. This in turn made it possible to define the major stages of development – Bronze Age, Iron Age, etc. – that form the framework of archaeology.

While archaeologists were using stratigraphy to study human prehistory, geologists were applying similar methods to the prehistory of the Earth itself. Between the mid-eighteenth century and the mid-nineteenth, they succeeded in establishing the relative age of the various rocks of which Earth's crust was formed; and they named and described the succession of geological eras – Cretaceous, Jurassic, etc. – through which the crust had passed. As they did so, it became clear that the changes that had taken place could not have occurred in a few thousand years, and they began to suspect that the rocks, and the animal and plant remains they contained, might be millions – or even hundreds of millions – of years old. But, like the archaeologists, they could prove only relative age, not absolute age, and it remained possible for opponents to ridicule their theories.

Between 1850 and 1950, the study of prehistory was transformed by the development of techniques that made it possible to put dates to events in the distant past, changing archaeology and geology from exercises in story-telling into scientific disciplines. Three examples of such techniques are tree ring analysis, radiocarbon dating, and uranium 238 dating.

Tree ring analysis, or dendrochronology, is the dating of wooden objects from the pattern of rings in the wood itself. In temperate regions, where there is a marked seasonal pattern to the year, trees stop making wood in winter. Each year's growth forms a 'ring'. A 70-year-old tree will display 70 rings in a cross-section of its trunk. In warm wet years, the rings are wider than in cool dry years. The rings therefore not only reveal the age of the tree, they contain a record of the variation in growing conditions from year to year. By studying the rings within long-lived trees still growing, it is possible to construct scales showing the sequence of favourable and unfavourable growing years in various regions, going back over hundreds, even thousands, of years. If the rings in the wood from a building or a boat can be matched to a known series, one can say where the timber grew, and when it was cut down. This enables scientists to make a good guess at when the object was made or built.

Radiocarbon dating is a technique used to ascertain the age of animal or plant remains up to about 40,000 years old. As animals and plants feed, they absorb carbon from their food, and this finds its way into the animal's bones or the plant's tissues (wood, in the case of a tree). Some of it is carbon 14, which is subject to continuous radioactive decay. When the creature dies, no further carbon is taken up into its tissues, but the carbon 14 continues to decay into other elements at a known rate. If a bone or a piece of wood is analysed, the level of radioactivity it displays enables testers to estimate the time that has elapsed since the animal or the tree died. For dates close to the present, the method is too unreliable to be of practical use; even for earlier dates, great care and skill are needed to obtain trustworthy results. But for objects many hundreds or thousands of years old, which have been handled and analysed with care, the margin of error can be kept within acceptable limits.

THE MEANING OF 'HALF-LIFE'   The half-life of a radioactive element is the time it takes for its radioactivity to fall by 50 per cent. Carbon 14 has a half-life of 5,700 years. A 5,700-year-old piece of bone or plant tissue will display 50 per cent of the radioactivity it would have displayed at the time of death. An 11,400 year-old specimen will register only 25 per cent. After 40,000 years the ratio will be down to 2 per cent. Uranium 238 decays in a similar fashion, but its half-life is a massive 4.5 billion years.

'Half-life' does not mean half the lifetime of a substance. Although the half-life of carbon 14 is 5,700 years, its radioactivity does not disappear completely in 11,400 years. After 11,400 years only a half of a half – in other words, a quarter – of its radioactivity still remains. As the following graph makes clear, significant levels of radioactivity remain long after twice the half-life has elapsed:

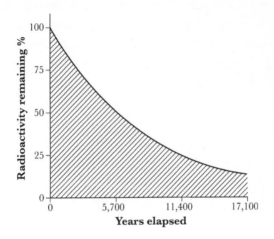

Figure 14. Half-life Curve of Carbon 14.

Uranium 238 dating works on the same principle. Uranium 238 is a radioactive variety of uranium – an isotope – that decays at a much slower rate than carbon 14. This makes it a perfect tool for measuring the age of rocks. The end product of the process of decay is lead. If a rock containing uranium is analysed, the ratio of lead to uranium 238 in the sample reveals how old the rock is. Using this, and similar techniques based on other radioactive minerals, geologists have put an age of 3.8 billion years to the oldest rocks in the Earth's crust. Samples from meteorites suggest an age of 4.5 billion years for the solar system as a whole.

THE GEOLOGICAL CALENDAR   Geologists divide the past into *eras*: the Palaeozoic, the Proterozoic, the Mesozoic, and the Cenozoic. These are subdivided into *periods*, and the most recent periods are subdivided into *epochs*. The oldest fossils date from about 3,000 million years ago. About 500 million years ago, in what has come to be called the Cambrian Explosion, there was a diversification of life into forms that were the precursors of the major divisions of the plant and animal kingdoms that we recognize today. The geological calendar since then is divided up as follows:

Period	Epoch	From (million years ago)	To	Life forms appearing
Cambrian		550	500	
Ordovician		500	440	Fishes
Silurian		440	410	Land plants
Devonian		410	360	Insects
Carboniferous		360	285	Reptiles
Permian		285	245	
Triassic		245	210	Dinosaurs, Mammals
Jurassic		210	145	Birds
Cretaceous		145	65	Flowering plants
Tertiary	Palaeocene	65	57	Horses
	Eocene	57	34	Monkeys
	Oligocene	34	23	Apes
	Miocene	23	5	Early hominids
	Pliocene	5	1.8	Early humans

Period	Epoch	From (million years ago)	To	Life forms appearing
Quaternary	Pleistocene	1.8	0.01	Modern humans
	Recent	0.01 (10,000 years ago)	Present	

NATURAL OR DESIGNED? In 1802, a book was published in England that contained one of the most powerful images ever evoked in a scientific argument. It was an image that captivated the book's readers, and it is one that still has power to persuade. The book was called *Natural Religion,* and its author was a clergyman named William Paley.

Paley invited his readers to conduct a thought experiment. Suppose, he said, that a traveller, crossing a deserted heath, came across a watch. He would know that it was not a natural object, but a *designed* one. And he would know that it must have had a *designer.* In the same way, said Paley, anyone coming across a flower or a butterfly, and observing their complex structure, and perfect adaptation to their way of life, must see them as created objects, and concede that they too had had a creator.

For most of Paley's readers, this was an argument that made nonsense of the idea that the wonderful adaptations seen in Nature could have arisen as a result of a process of so-called 'evolution'. Paley's thesis, and the image in which it was embodied, held the field virtually unchallenged for 60 years after his book was published. But in 1859, another man met his argument head-on, showing that this appearance of perfect adaptation could just as easily be explained by a process of step-by-step evolution of exactly the kind that Paley had set out to ridicule.

DARWIN AND NATURAL SELECTION The man who refuted Paley's argument, and created the revolution that determined the future course of biology, was an English doctor's son named Charles Darwin, who was born in the market town of Shrewsbury in 1809. In the autumn of 1828, as a 19-year-old student in Cambridge, he moved into the very college rooms that Paley had occupied 70 years earlier. Paley's book was a set text on Darwin's course, and he was entranced by its arguments. On his graduation, at the age of 22, he sailed as a naturalist on a round-the-world naval survey. The observations he made during the voyage initiated a train of thought that caused him to change his views, and finally led him to a diametrically opposite conclusion.

In South America, Darwin discovered remains of extinct creatures that seemed to bear a *family resemblance* to living animals in the same locations. On the islands of the Galapagos, he found different species of birds that, again, seemed to bear a family resemblance to one another, but which bore no resemblance to birds in similar environments in other parts of the world. Paley had invoked the perfect adaptation of plants and animals to their circumstances and way of life as evidence of divine creation. But on this argument, one would have expected to find similar species in similar environments all over the world. What struck Darwin was the enormous *difference* in living forms in similar but widely separated locations.

In 1838, two years after his return, he saw an orang-utan at a zoo and realized that apes and human beings too displayed what seemed to be family likenesses. Between the ages of 28 and 34, he conducted an extensive programme of reading and reflection; by 1844 he had effectively completed his theory of *evolution by natural selection*. In an unpublished essay written in that year, he explained the marvellous variety of the natural world as evidence of a long process of evolution under the pressure of environmental factors and the struggle for survival between, and within, competing species.

In the climate of the time, it was a dangerous piece of work for a respectable gentleman to put his name to; so he wrapped the essay in brown paper, and hid it in a cupboard. He placed with it a letter addressed to his wife, instructing her to have it published after he was dead. Had his hand not been forced 15 years later, by a letter from another naturalist proposing the same idea, his speculations would probably have remained hidden from the world until after he had left it.

TWO VIEWS OF EVOLUTION Darwin was not the first person to put forward the idea of evolution. His grandfather, Erasmus Darwin, had published a long poem about it. The French naturalist Lamarck had written a book, *The Philosophy of Zoology,* in which he argued forcefully that present-day species had arisen by a process of evolution from simpler ancestral forms. But Darwin was the first person able to suggest a mechanism by which it might have taken place: one that was backed up by detailed argument and supported by a vast quantity of evidence. And there was a crucial difference between his view and that of other evolutionists.

Before Darwin, the image employed by evolutionists was that of a ladder that led up from ancient, simpler, life forms to a topmost rung, where modern man stood in solitary splendour. The image that encapsulated Darwin's view of evolution was utterly different. It involved no concept of 'lower' or 'higher' forms of life. It was a tree, in which the spreading branches were the living and extinct forms of life that had evolved from a common ancestry. On his tree, humanity was just a twig, on a branch that had no more significance than any other branch on the spreading tree of life.

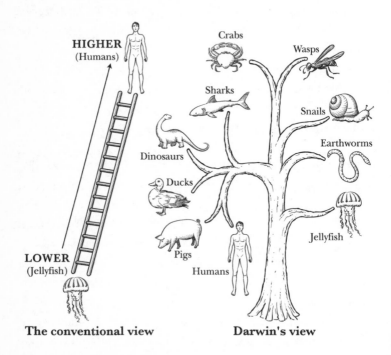

The conventional view

Darwin's view

Figure 15. Two Views of Evolution
Most evolutionists saw humanity as the culmination of an upward progress. Darwin saw humanity as a twig on the spreading tree of evolution.

DARWIN'S EXPLANATION Darwin's contribution to biology was not the idea of evolution: it was the discovery of the process by which evolution occurs. He looked at the similarities between living and extinct forms of life, and between living species in adjoining locations, and he saw them as evidence of a shared ancestry. He studied the work of plant and animal breeders, and observed how they were able to exploit inheritable differences between members of the same species to create new varieties. He called this *artificial selection*. He guessed that a similar process of selection occurred in nature; and that this could, over a long period of time, give rise to totally new species. In artificial selection, the selecting agent was the hand of the stockbreeder. In nature, it was the struggle for life: the competition between, and within, species for the means of existence. He called this process *natural selection*.

Darwin's explanation was different from Lamarck's. Lamarck thought evolution was a result of the passing-on of acquired characteristics. Land-based animals, in other words, learned to swim, and their descendants inherited this *learned* skill, and became aquatic. Darwin's view was that, if conditions changed in a way which favoured animals that were innately better at managing in the water, they would survive, and pass on this *innate* skill to their offspring, leading in time to a species that was collectively better adapted to life in the water.

ANIMAL RELATIONSHIPS Not all likenesses are family likenesses. Take hedgehogs. A snout at the front, a tail at the back; always grubbing in the earth. And they snuffle and squeal, just like pigs. They look as if they might be related to pigs.

They are about as closely related as mice are to gorillas. Whoever christened them 'hedgehogs' was wide of the mark.

Hedgehogs are Insectivora (insect-eaters). They are only distant relatives of pigs. If a pig wants to meet a close relative, it has a better chance of doing so on the beach than under a hedge: pigs and porpoises are cousins under the skin. Porpoises are Cetaceans (aquatic mammals), an order closely related to the Artiodactyla (hooved mammals), the order to which pigs and sheep belong. Fossils of four-footed whales from 50 million years ago show how close the relationship is.

It's all down to evolution. When the ancestors of the porpoises ventured into the sea, they said goodbye to their relatives who remained on land – the ancestors of the sheep and the pigs. Under the pressure of natural selection, the two lines diverged, so that only a trained zoologist would now think of grouping them together. But if you see a porpoise, feel free to call it a 'seahog'. Science is on your side.

THE MARSUPIAL TASMANIAN WOLF   A striking example of deceptive appearances is the Tasmanian wolf, or thylacine. Although it looked like a wolf, it was not even a placental mammal. It was a marsupial, like the kangaroo. Once again, the confusion was the result of evolution. Pigs and porpoises evolved from a common ancestor as a result of *divergence*: a change in form in response to differing environments. Wolves and thylacines are examples of *convergence*: animals with very different pedigrees that have evolved outwardly similar forms as a result of adaptation to similar environments.

Unfortunately, you won't ever see a thylacine. Nineteenth-century thylacines made the mistake of developing a taste for sheep, and a price was put on their heads. The last one died in a zoo in 1936.

ALFRED RUSSEL WALLACE The Englishman Alfred Russel Wallace was one of the nineteenth century's greatest naturalists. He came from a modest background, and was initially apprenticed to a land surveyor. In his twenties he went on an expedition to South America, to hunt for specimens to sell to collectors back in England. During 4 years in the jungle, he amassed a superb collection; but lost it when the ship carrying it sank in a storm. Undeterred, he travelled to the East Indies, and began collecting all over again. While he was there, in 1858, he independently hit upon the idea of evolution by natural selection that Darwin had been nursing for 20 years. Knowing Darwin's reputation, but unaware of the extent of his theorizing, he wrote to him to tell him of his speculations. When Darwin revealed the nature of his own theory, and the evidence he had accumulated in support of it, Wallace yielded priority; for the rest of his life he was Darwin's devoted publicist. His own reputation was ensured by *The Malay Archipelago* (1869), the book he published on his return.

The loss of his South American specimens was just one of several misfortunes Wallace suffered during his career. In 1870 he had the bad luck to get involved with a Flat-Earther named John Hampden. Hampden had published an advertisement in which he offered to pay up to £500 (£30,000 in today's money) to anyone who could prove the Earth was round, by 'exhibiting … a convex river, canal, or lake.' The condition was that the person accepting the wager would forfeit an equal sum if the experiment were unsuccessful. Wallace, confident of success, took up the challenge.

The experiment was carried out on a 9.6-kilometre/6-mile-long straight stretch of a canal called the Bedford Level. Wallace had calculated, from the known dimensions of the Earth, that if three markers were set up at identical heights above the water at 4.8-kilometre/3-mile intervals on bridges over the canal, the middle one, when viewed through a telescope, would appear to be 1.5 metres/5 feet higher than the other two. Hampden's referee – a Flat-Earther himself – looked through the telescope and brazenly declared that the three markers were perfectly in line. However, Wallace's referee confirmed that the result was as Wallace had predicted, and declared him the winner. There followed a long and vicious dispute, in the course of which Hampden was imprisoned for criminal libel. Although Wallace's science was sound, and his proof was immaculate, the case did his reputation no good. In the end, he got to keep the £500. But he incurred heavy costs, which he was prevented from recovering by Hampden's bankruptcy.

MENDEL'S EXPERIMENTS The city of Brno (formerly Brunn), the second city of the present-day Czech Republic, was the home of the Bren gun, to which it gave its name. In the mid-nineteenth century it was the home of a man who gave it an even greater claim to fame, as the scene of one of the most important experiments in the history of biology. His name was Gregor Mendel, and he was a monk in the city's St Thomas Monastery. Historians have sometimes written as if the monastery were a rural backwater in an unknown land, rather than what it truly was: a centre of culture and learning in a proud, historic, and fast-growing industrial city of 70,000 people.

Mendel, a bright boy from a poor farming family, was born in 1822 in a village called Heinzendorf, in the German-speaking region of Silesia. He went from his village school to the local high school, and then to the Philosophical Institute in the nearby city of Olmutz (now Olomouc, in the Czech Republic). But poverty prevented him from going to university, and in 1843 he became a monk on the recommendation of one of his professors. The monastery was an Augustinian foundation, dedicated to teaching, and the monks provided the mathematical instruction at the Philosophical Institute.

Mendel was well grounded in science, especially in physics and mathematics, and the monastery routine gave him the time to pursue his own research. He decided to study the consequences of cross-breeding, and he began with mice, whose rapid reproduction made it easy to study the characteristics of successive generations. But when it was put to him that studying the sexual activities of small mammals was perhaps not a suitable occupation for a monk, he transferred his attention to the monastery garden. His abbot gave him a plot of ground 30 metres long by 7 metres wide/100 feet by 20 feet, and a greenhouse, and in 1855 Mendel resumed his researches via the less stimulating medium of garden peas. He selected 22 varieties and, in a carefully planned and rigorously controlled experiment, he analysed the inheritance of 7 separate characteristics through successive generations.

Fourteen years - and 300,000 peas – later, he was appointed abbot, and the day-to-day responsibilities put an end to his detailed researches. His work brought him no recognition in his lifetime. He died in 1884, loved and honoured in his native town, but unknown to the wider world. Mendel had read, and closely annotated Darwin's *The Origin of Species*; but Darwin died just 18 months after Mendel, without knowing that the key to the question he had never been able to answer – how hereditary characteristics were passed from generation to generation – had been lying for 17 years in a Czech monastery's files. Even if he had known of it, it is questionable whether he would have recognized the significance of it for his own theory. Darwin was a poor mathematician; and the form in which Mendel's results were presented would not have appealed to his cast of mind.

Mendel's work was not rediscovered until the early twentieth century; it was not until the 1930s that the work of these two great biologists was brought together in the synthesis called neo-Darwinism, out of which our present-day understanding of evolution was born.

INHERITED CHARACTERISTICS Many scientific experiments are conducted in order to test an idea that has already occurred to the scientist, but Mendel's researches were driven by curiosity, not by a preconception. He just wanted to understand the mechanics of inheritance. In choosing peas, he was luckier than he knew. The characteristics he studied – green versus yellow colour, wrinkled versus smooth skin, and so on – are either/or qualities. The kind of peas he grew *are* either green or yellow, and either smooth or wrinkled: they don't come in greeny-yellow varieties, or have skin that is halfway between wrinkly and smooth. The either/or nature of these characteristics made it easy to track their inheritance through successive generations, and to subject that inheritance to statistical analysis.

Mendel applied the pollen himself, taking care to ensure that none of his plants were fertilized by insects. He kept careful records of the plants from which the pollen had been collected, and to which it had been transferred. In this way he knew which plants were the parents of which seedlings. What he discovered – to take yellow and green colour as an example – was that when he crossed a yellow pea with a yellow pea, or a green pea with a green pea, the offspring were always the same colour as the parent. But when he crossed a yellow pea with a green pea, all the offspring were yellow. But if he crossed two yellow peas from this generation of mixed parentage, a quarter of their offspring were green. This proved that the inherited characteristics did not get mixed up, like green and yellow paint, but were passed down the generations intact. At the time

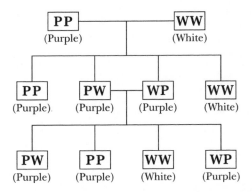

Figure 16. Flower Colour and Mendelian Inheritance
A plant carrying only genes for purple colour is crossed with one carrying only genes for white. If purple colour is dominant over white, three-quarters of the offspring will have purple flowers. If two purple-flowered plants from the second generation have inherited both genes, one-quarter of their joint offspring are likely to have white flowers.

Mendel was conducting his researches, the word *gene* did not exist. In modern terms, we would say that different genes for colour were being passed on. The gene for yellow colour was *dominant* over the gene for green, concealing the fact that some of the mixed-parentage plants were carrying the gene for green colour, which could reappear in a later generation if a plant received a green colour gene from both parents.

Mendel's work had two important outcomes. Firstly, it yielded a mass of statistical evidence that would in due course form the foundation of the science of genetics. Secondly, it illustrated two concepts that would become central to that science: the *phenotype*, or outward appearance of an animal or plant, and the *genotype*, the inherited and transmissible factors that lay beneath that appearance.

PASTEURIZATION After Jenner's demonstration of the effectiveness of vaccination in controlling smallpox, it might have been thought that similar treatments for other diseases would quickly have followed. But it was not until a 180 years after Jenner's first vaccination that the next advance was made, by a French chemist called Louis Pasteur. Pasteur, the son of a tanner, was born in Dôle, in the Jura, in December 1822. As a youth, he was not a promising student. In so far as he showed any aptitude at all, it was for painting; his early ambition was to be a professor of fine art. In 1842, after a spell as a teacher in the Royal College in Besançon, he obtained his *baccalauréat* in science, but with the grade '*médiocre*' attached to his performance in chemistry. In 1843, when he was 21, he took a high place in the list of admissions to the École Normale in Paris. Like many famous scientists, his course in life was determined by the influence of an inspiring teacher, in his case the chemist Antoine Jerôme Balard. In spite of his inauspicious start in the subject, Pasteur was determined to make chemistry his life's work. Within a few years, the erstwhile 'mediocre' student had conducted a programme of research that gained him international celebrity. His investigations into the optical characteristics of crystals of organic substances, completed when he was 26, gained him the Royal Society's Rumford Medal.

In 1854 he was appointed dean of the Faculty of Sciences at Lille University, where he became interested in the problems of the wine industry, which was losing huge sums of money as a result of deterioration in its stocks. Many of his contemporaries, including the eminent German chemist Justus von Liebig, insisted that fermentation was a chemical process that did not require the involvement of any living organism. With the aid of the microscope, Pasteur discovered that there were actually two such organisms – both varieties of yeast – that were the key to the process. One produced alcohol. The other produced lactic acid, turning the wine sour. To ensure that the bad kind did not remain in the wine, Pasteur recommended that it be killed by gentle heating to 44°C/120°F. Despite the industry's initial horror at the idea of heating wine, a controlled experiment with heated and unheated batches proved the effectiveness of the procedure. *Pasteurization* – the process that now ensures the safety of so many of the world's food products – had been born.

SPONTANEOUS GENERATION   Pasteur's work on fermentation enabled him to bring to an end one of the longest-running disputes in the history of biology: the question of 'spontaneous generation'. Many biologists held that it was possible for microscopic life to appear, in effect, out of thin air, or at least out of dead matter. Pasteur first established that the dust in the air contained spores that would germinate if introduced into a nutrient broth. He then devised an experiment in which boiled-up broth was exposed in a flask with a bend in its neck, which admitted air, but which dust particles could not negotiate. This time, no living organisms developed in the flask, and no decay occurred. He announced his results at a glittering gala at the Sorbonne in 1864. It was a triumph. The theory of spontaneous generation was history, and the science of bacteriology had been well and truly launched.

THE DISCOVERY OF INFECTION Having solved the problems of the wine industry so brilliantly, it was natural that when a disease of silkworms began ruining the silk industry in the south of France, Pasteur should again be called in. Pasteur knew nothing about silkworms, but when he turned his microscope on the problem, he identified a minute parasite infesting both the worms and the mulberry leaves on which they were fed. His prescription was drastic: the infected worms and shrubs must be destroyed and replaced with new stock. His reputation ensured that his advice was followed, and another industry was saved.

His work with the silkworm disease turned his attention to the wider subject of communicable disease in general. Traditional medical thought, going back to the Greeks, insisted that disease was an affliction of the individual organism. The idea that diseases might be passed around between living creatures was anathema. It was another of those blind spots that seem so inexplicable to us now, brought up as we are with the fear of germs and infection from our earliest childhood. However, there was nothing *obvious* about the idea of a communicable disease. The fact that people contracted a disease at the same time did not of itself prove that they got it from one another. The fact that bacteria were found in wounds, or in sick people, was not in itself evidence that the bacteria were the cause of the problem. When Pasteur suggested that it might help if doctors and nurses were to wash more frequently, sterilize their instruments, and steam their bandages, it wasn't just wounded self-esteem that caused their hackles to rise, it was a genuine sense of outrage at what at first seemed an outlandish theory.

Pasteur's politics were very right wing, and it has been suggested that his preoccupation with the threat to the human body of millions of little microbes was not wholly unconnected with his equally strong conviction of the threat to the body politic of the swarming masses. But whatever the inspiration for his germ theory, it remains, in the opinion of many good judges, the most important theoretical development in the history of medicine. And it was but one of a host of important discoveries that we owe to this genius of biochemistry.

PASTEUR'S VACCINES   In the autumn of his life, Pasteur addressed himself to the possibilities of vaccination, first explored by Jenner. He began by tackling the deadly animal disease, anthrax. He found that by heating a preparation of anthrax germs, he could reduce their virulence, and produce a mild version of the disease that did not kill the animal inoculated, but was still able to provoke an immune response. In 1881 he provided a dramatic demonstration of the efficacy of his vaccine by having half a flock of sheep injected with it. He then injected both this half of the flock and the unprotected half with the disease. All the unprotected sheep died. Every vaccinated animal survived.

He next developed a vaccine that he hoped would prove effective against rabies, and it was with this vaccine that he performed his last, dramatic experiment. In 1885, when he was 62 years of age, a boy was brought to him who had been badly mauled and bitten by a mad dog. He administered the as-yet-untested vaccine to the boy, and saved his life. It was a fitting climax to the amazing career of a man who deserves to stand alongside Aristotle and Darwin in biology's Hall of Fame.

MAURY AND OCEANOGRAPHY   By the 1840s, the science of geology – the study of the structure of the Earth, and the history of its rocks – had well and truly come of age. But the science of oceanography – the study of the seas and *their* history – had not yet been born. The principal reason was money. Geology was ideally suited to the gentleman amateur, with time to wander the countryside, hammer in hand, and with access to a circle of like-minded friends with whom to discuss his speculations. The study of the oceans required costly expeditions; without finance from governments or wealthy patrons, worthwhile results could not be achieved. That the science *was* created in the 1850s is largely down to one man: the American Matthew Fontaine Maury.

Maury was born in Fredericksburg, Virginia, in 1806. He was the son of a farmer, and he joined the navy as a midshipman. By the time he was 24, he had completed a circumnavigation of the globe. In 1839, he was made lame by a stagecoach accident, and forced to resign his commission. He was appointed to what could have been an undemanding post: superintendent of the Depot of Charts and Instruments. But he threw himself with tireless energy into the performance of the office, and in particular to the study of winds and currents. To aid the collection of data, he designed specially printed logbooks, which he distributed to ships' captains.

Maury was convinced that the study of the oceans could not be pursued without cooperation on the part of the maritime nations. Largely as a result of his efforts, an International Conference was held in Brussels in 1853, at which governments agreed to adopt a standardized system of weather recording. In 1855, he published the world's first textbook on oceanography, *The Physical Geography of the Sea*.

On the outbreak of war in 1863, he was appointed head of coast defences for the Confederacy. After the defeat of the Confederacy, he went into exile, first in Mexico and later in England. In 1868, he returned to the United States, and was appointed professor of physics at the Virginia Military Institute, where he spent the last five years of his life. In 1930, his creation of the science of oceanography was recognized by his election to the Hall of Fame for Great Americans.

SURVEYS AND EXPEDITIONS In the quarter-century following the publication of *The Physical Geography of the Sea*, there was an explosion of knowledge of the deep ocean: its configuration, its currents, its weather, and its life forms. An important factor in this expansion of knowledge was the laying of underwater cables. The first such cable was laid under the Straits of Dover in 1851. The laying of cables called for a better knowledge of the ocean bed, of its currents, and of its temperature variations. It encouraged the development of new technologies, which proved invaluable in submarine surveys. And the cable-laying companies had the money to make such surveys possible.

The opportunities for research opened up by cable-laying were at first not recognized by marine biologists, because of a widespread belief that life could not exist in the conditions of extreme cold, pressure, and darkness prevailing in the ocean depths. But during the 1860s, evidence began to accumulate that this assumption was unjustified. In 1868, two British biologists, W.B. Carpenter and C. Wyville Thompson, recognizing the benefits that might accrue from a properly equipped expedition to investigate the physical conditions and life forms of the world's oceans, persuaded the Royal Society to support a series of expeditions to dredge the ocean depths. In the course of two voyages, first in the *Lightning*, and then in the *Porcupine*, they collected incontrovertible evidence that the oceans were populated by a vast array of life forms hitherto unknown to science.

As these expeditions proceeded, Carpenter became more and more intrigued by the data the ships were accumulating concerning temperature, density, and the movement of currents at various depths. He became convinced that there was a persisting pattern of deep-sea currents within the oceans, and that this circulation was a major factor in climate change, including the causation of ice ages. This proposition met fierce opposition from the leading British theorist of ice age causation, the Scotsman James Croll, who believed that the main driver of climate change was the trade winds, and that these were the cause of changes in the behaviour of ocean currents. Carpenter was convinced that his theory would be vindicated if a third, more ambitious, survey could be mounted in hitherto unexplored oceans. With the support of the Royal Society, he persuaded the British Government to commission a worldwide ocean survey by the naval corvette HMS *Challenger.*

Between 1872 and 1876, with Thompson on board as chief scientist, the ship crisscrossed the oceans of the world. The results, which were published in 50 volumes between 1880 and 1895, were a landmark in oceanography; they established beyond doubt that there was a persisting pattern of circulation of sea water around the globe, of exactly the kind that Carpenter had suggested.

THE DISCOVERY OF X-RAYS Some important advances in science have been the result of serendipity – a chance discovery while researching something else. A good example is the discovery that was made in a laboratory in the University of Würzburg in Germany in November 1895.

Wilhelm Conrad Röntgen was born in Lennep, in Rhenish Prussia, in 1845. The son of a textile merchant, he was educated in Holland and Switzerland, and trained to be a mechanical engineer, but he changed his mind, and decided to make a career in physics. Between the ages of 30 and 50 he held a number of university posts, including professorships in Strasbourg, Munich, and Würzburg.

On 8 November 1895, he was experimenting with cathode rays – negatively charged particles emitted by an electrode in a discharge tube – which were at that time little understood. He suddenly noticed that a screen more than a metre away from his apparatus had become illuminated in a quite unexpected way. The screen was coated with a substance called barium platinocyanide, and Röntgen realized that it was too far away for the effect to have been caused by the cathode rays. It occurred to him that some hitherto unknown form of radiation might have been emitted by the tube. During the following month, with increasing excitement, he made a succession of surprising discoveries concerning the properties of this strange radiation. He established that it was not deflected by a magnetic field. He also found that it was capable of passing through many solid materials, including wood and metal, and his wife's hand. When a photographic plate was placed behind a hand, it produced an image, not of the hand, but of the bones within the hand. He announced his discovery on 28 December, giving his mysterious rays the appropriate name of *X-rays*.

This radiation had other surprising properties. It behaved like visible light, but it was not reflected or refracted in the way that light was. The medical applications were obvious, and X-ray photography quickly became an accepted part of medical diagnosis. But the true nature of the radiation did not become clear until 1912, when another German physicist, Max Theodor Felix von Laue, established that it was a form of electromagnetic radiation of a much shorter wavelength than visible light.

At this time, there was no understanding of the effect that exposure to such radiation could have, and both Röntgen and his assistants suffered as a consequence. But he had the consolation of being awarded the first ever Nobel Prize for physics.

BECQUEREL'S DISCOVERY Röntgen's chance discovery was echoed by the experience of another physicist, who had decided to investigate X-rays further. The lucky person in this case was a Parisian academic named Antoine Henri Becquerel. Becquerel, who was born in 1852, was seven years younger than Röntgen. He came from a family of physicists, and in 1891 he took up a post in the Museum of Natural History in Paris that his father and grandfather had held before him. In 1895 he became professor of physics at the École Polytechnique.

Becquerel's father had done important research into the phenomenon of fluorescence; his son had maintained this interest. Having studied Röntgen's results, it occurred to him that there might be fluorescent materials that emitted X-rays. In February 1896, he wrapped some photographic film in thick, black paper, and left it in sunlight, with a crystal of a fluorescent material – potassium uranyl sulphate – on top of it. His hope was that, when the sunlight caused the crystal to fluoresce, any X-rays that might be emitted would penetrate the black paper and record an impression on the film. To his delight, when the film was developed, it showed fogging of the kind produced by X-rays. He tried to repeat the experiment, placing a copper cross between the crystal and the film, but was frustrated by a succession of cloudy days, and he eventually decided to develop his film anyway. To his amazement, it contained an image of the cross, indicating that the radiation responsible had its source within the crystal itself, rather than being the product of fluorescence induced by sunlight. This was an astounding discovery, as there was no source of energy known that could simply emanate from a solid material.

Further investigation showed that, whatever the radiation was, it was not X-rays, since it was capable of being deflected by a magnetic field. This meant that it must be composed of charged particles; but what reaction was capable of producing them remained a mystery – for the time being.

PIERRE AND MARIE CURIE   The answer to the riddle of Becquerel's mystery radiation was found by Pierre and Marie Curie, the most successful husband-and-wife team in the history of science. Marie was born Marya Sklodowska, in the Polish capital city of Warsaw, in 1867. Her father, Wladyslaw, was a science professor who had been prevented from teaching because of his involvement in a failed uprising against Russian occupation four years before she was born. She was a brilliant high school student, but was unable to obtain a university education in her native country.

Through hard work and self-denial, she saved enough to get her to Paris, and to support herself while she studied at the Sorbonne. In 1893, as the university's first-ever female graduate in physics, she was placed first in the examinations for the *licence* (degree) in physical sciences, and in 1894 she followed up this success with second place in mathematical sciences. It was in that year that she met Pierre Curie, a teacher in the university, eight years her senior, who was studying the electrical properties of crystals. They were married in 1895, and embarked on a joint programme of research that was to have profound consequences, not only for the study of physics, but for chemistry and medicine as well.

Pierre's gifts – as an experimentalist, as a theorist, and as a designer of delicate instruments – were crucial to the success of these investigations; but, as he himself recognized, he was married to a genius, whose lead he was happy to follow. In 1897,

Marie gave birth to a daughter, Irene. She had continued to work throughout her pregnancy, and she and Pierre were able to continue with their research afterwards, thanks to the devoted childcare provided by Pierre's widowed father.

Within weeks of her baby's birth, Marie began working for a doctoral thesis, for which she chose to study the radiation discovered by Becquerel in the previous year. She quickly established that it was produced by uranium – one of the elements present in Becquerel's crystal – and that the level of radiation depended solely upon the *quantity* of uranium present, and was unaffected by sunlight, temperature, or the chemical condition of the uranium. The implication seemed to be that it was a totally unknown phenomenon, and a property of the uranium atoms themselves.

The next question they addressed was: was uranium the only element that possessed this property? Before long they had discovered that another element, thorium, behaved in exactly the same way. The new phenomenon needed a name: Marie christened it *radioactivity*. She next examined various ores of uranium and thorium, and discovered to her amazement that they were *more* radioactive than the elements themselves. This could only mean one thing: they were dealing with a highly radioactive element unknown to science.

They now decided to concentrate their attention on the mineral pitchblende. As they worked to separate the various elements in the ore, they realized that it contained two radioactive sources in addition to uranium. They had discovered not one, but two new elements. They named one *polonium,* in honour of Marie's native country, and the other *radium,* from the Latin word for a ray of light.

They had complete faith in the existence of their two new elements, but many of their scientific peers were unconvinced. To prove their case, it was necessary to produce an acceptable quantity of the elements in their pure state. There was a cheap source of the raw material they needed in the form of waste from the Bohemian glass industry. The separation of the minute traces of the radioactive elements from the tons of glass they had to import and process took them four years of grinding toil in appalling conditions. But at the end of the four years, in 1902, they had $\frac{1}{10}$th of a gram of pure radium. The doubters were silenced at last.

In 1903, their achievement was recognized by the joint award to them and Becquerel of the Nobel Prize for physics, for their work on radioactivity. In 1904 Pierre was appointed professor of physics at the Sorbonne. Two years later he died in a road accident. Marie was devastated, but determined to continue their work. She was appointed to the professorship in her husband's place, becoming the first woman teacher in the university. In 1911 she became the first person to win a second Nobel Prize – a feat not equalled for another 50 years – when she was awarded the prize for chemistry for her achievement in isolating pure radium. She died in 1934, at the height of her fame, of leukaemia caused by the radioactivity she had devoted her life to studying.

THE DISCOVERY OF THE ELECTRON As the nineteenth century progressed, the difference between atoms and molecules, and the way atoms joined together to form molecules, was gradually teased out. But for much of the century, Dalton's image of the atom as the irreducible building block of matter remained the guiding idea of both chemistry and physics. As the century drew to a close, this image was destroyed by the discovery that atoms were composed of much smaller units, and that what had been thought of as solid lumps of matter were mostly empty space. This transformation was in large part due to the work of two people: Joseph John Thomson and Ernest Rutherford.

'J.J.' Thomson was born near Manchester in 1856. He was originally marked out for an engineer. But his father's death meant that he could not afford an apprenticeship. He had attended the local Owens College in his early teens, and became interested in physics. He formed an ambition to go Cambridge, and won a maths scholarship to Trinity, the college where Newton had once lived, and Maxwell still did. He did well in his examinations, and was made a university lecturer.

In 1884, when he was 27, he was made director of the university's Cavendish Laboratory, where he remained for 35 years. Under his leadership, it became the leading institution of its kind in the world. Part of the explanation for its success was the availability of funds from the Great Exhibition of 1851 to finance the recruitment of outstanding researchers.

Thomson's own outstanding achievement was the discovery of the *electron*: a negatively charged particle with only $\frac{1}{2,000}$th of the mass of a hydrogen atom. It won him the 1906 Nobel Prize for physics. He insisted that electrons were a fundamental part of the structure of atoms, which he visualized as hard spheres, in the outside of which the negatively charged electrons were embedded, like fruit in a cake, in just sufficient numbers to neutralize the atom's positive charge.

ERNEST RUTHERFORD   In 1895, Thomson was joined by an assistant who had won one of the Great Exhibition scholarships, a New Zealander by the name of Ernest Rutherford.

Rutherford had been born in Brightwater, New Zealand, in 1871, the second of 12 children. He showed great promise at school, and won a scholarship to Canterbury University. He had actually come second in the local Great Exhibition scholarship competition, but the winner had decided to remain in New Zealand. The story goes that he was in the fields digging potatoes when the news arrived, and that he threw down his spade, saying, 'That's the last potato I'll dig!'

Rutherford assisted Thomson for two years, and then applied for the post of professor of physics at McGill University in Canada. In 1898 arrived in Canada to find himself in charge of probably the best-equipped laboratory in the western hemisphere, financed by a wealthy tobacco manufacturer.

PROTONS AND NEUTRONS   Rutherford's interest in all types of radiation had been greatly stimulated by hearing of Becquerel's discovery of radioactivity while he was working in Thomson's laboratory. As soon as he had settled in his new post he began a programme of research into radioactivity that would occupy him for 40 years. In 1904 he wrote *Radioactivity,* the first textbook on the subject, which became an instant classic.

As his fame spread, other universities tried to tempt him away from McGill. In 1907, the professor of physics at Manchester University offered to relinquish his chair, on condition Rutherford would accept it. Rutherford agreed, and spent the next 14 years there. The Manchester laboratory already had a fine reputation and, under his leadership, it attained an eminence second only to the Cavendish in Cambridge.

Between 1907 and 1909, Rutherford and his assistant Hans Geiger (the inventor of the Geiger counter for measuring radioactivity) conducted research into the nature of alpha particles: the positively charged particles emitted by certain radioactive elements. Among other experiments, they fired particles at extremely thin sheets of gold foil. Almost every one – about 7,999 out of every 8,000 – passed through without being deflected. This led Rutherford to conclude that the atoms of gold, and of other elements, were made up of a vast expanse of empty space, with a minute, hard core. In 1919, when he was once more back in Cambridge, he conducted an experiment in which he used alpha particles to bombard nitrogen. This caused the gas to emit positively charged particles, which Rutherford was able to show were present in all atomic nuclei. He named these particles *protons*.

A few years later, James Chadwick, a New Zealand scholarship winner who had worked with Rutherford in Manchester, pointed out that atomic nuclei could not be composed solely of protons. If they had enough protons to account for their atomic weight, he argued, atoms would have too large a positive charge. In 1932, he proved the existence in the nucleus of a neutral (non-electrically charged) particle, of approximately equal mass to the proton, which named the *neutron*.

As a result of the work of Thomson, Rutherford and Chadwick – and the many other able scientists who added to their discoveries – the world at last had a meaningful and workable picture of the structure of the atom:

1. A compact nucleus, where almost all of the atom's mass is located, made up of positively charged protons and electrically neutral neutrons

surrounded by

2. A vast (atomically speaking) expanse of space in which the negatively charged, and almost weightless, electrons navigate their orbits.

For a while, around 1920, this *planetary model of the atom* seemed to contain the essential truth of the structure of matter. It was a brief honeymoon period. Within a few years, new discoveries – and a strange and difficult new science called *quantum mechanics* – revealed that it was a gross oversimplification of the underlying reality. But for the non-mathematician, it remains the best available representation of the stuff of which we – and the world – are made.

IMPROVING ON MENDELEYEV There is a danger, when writing about chemistry, of giving Mendeleyev more than his due. His was certainly a great achievement, but he was very much building on foundations that had been laid by those who came before him. And the modern version of the periodic table (*see page 170*) owes much to those who came after him, in particular the English physicist Henry (Harry) Moseley.

Moseley was born in the English seaside town of Weymouth in 1887, into a family steeped in science. His father had been a professor of anatomy, but had died when his son was four years old. Both his grandfathers had been distinguished scientists, and his elder sister did important work in biology. Moseley himself won scholarships to both Eton College and Oxford. On his graduation in 1910, he joined the brilliant young team at Rutherford's laboratory in Manchester, where he twice had the good fortune to meet – and exchange ideas with – the Danish physicist Neils Bohr. In the few years he worked there he did work of surpassing importance on the atomic structure of the elements, using the recently developed technique of X-ray diffraction analysis.

Mendeleyev had arranged his elements in ascending order of their atomic *weight*. This left some uncertainties unresolved. He had no way of knowing whether there was a minimum interval between the weights of the various elements, and so he could not be certain how many unknown elements might exist between any two in his table. The atomic weight itself was an uncertain quantity, owing to the presence of what we now know are *isotopes* – different versions – of some of the elements. Moseley swept away these uncertainties, and put the periodic table on a solid footing. As a result of his work on the electrical charge of the atomic nucleus, he was able to propose a superior ordering of the elements – not by atomic weight, but by what he called atomic *number.* This was a measure of the positive charge on the nucleus, which we now recognize as the number of protons in the nucleus. Because this was always a whole number, it was possible to say how many natural elements there must be (including elements still to be discovered) between the simplest (hydrogen) and the most complex then known (uranium). In 1912, at the age of 26, Moseley published his law of atomic numbers, with a new version of the periodic table, which was essentially – up to element 92 – the table set out in figure 13. It would undoubtedly have earned him a Nobel Prize, had he lived to be awarded one. But when war broke out in 1914, he volunteered at once; 12 months later – aged 28 – he was killed at Gallipoli.

MENDELEYEV IN LONDON Round about the time the young Moseley went up to Oxford, the 70-year-old Mendeleyev paid a visit to the Royal Institution in London, to give a lecture, and to receive the Chemical Society's Faraday Award. It was customary to present the recipient with a silk purse containing gold sovereigns. Mendeleyev politely emptied his on the table, and took just the purse, saying that nothing would induce him to accept money for speaking in Faraday's memory, in a place made sacred by his work.

THE DRIFTING CONTINENTS Anyone who looks at a map of the Atlantic is likely to notice that the outline of the eastern coasts of North and South America seems to follow that of Western Europe and the west coast of Africa. The correspondence was commented upon almost as soon as the first maps of the Americas were produced in the sixteenth century; but few people thought it was anything more than a coincidence.

In 1911, a German meteorologist and explorer named Alfred Wegener came across a paper that drew attention to another striking correspondence: between the fossils in the rock formations of West Africa and those of Brazil. The writer of the paper suggested that there might once have been a land bridge between the two continents. Wegener suggested an even more startling possibility: that the continents had once formed a single landmass, but had drifted apart.

Wegener's theory was that the continents were masses of light rock, floating in the heavier rock that made up the ocean floor, and that they had forced their way through these rocks, like an ice-cutter slicing through a frozen sea. But he was unable to suggest a process that could cause the continents to move around in this way, and his theory of 'continental drift' was dismissed as far-fetched. He died on an Arctic expedition in 1930, leaving behind a world unconvinced of the value of his theory.

In that same year, a British geologist, Arthur Holmes, suggested a mechanism that might be driving the process Wegener had described. Holmes' thesis was, that convection currents in the Earth's core, driven by the heat from radioactive decay, could be carrying the continents around the globe. But like Wegener, Holmes was unable to convince the scientific community, and the theory of continental drift remained a minority interest for another 30 years.

CRACKS IN THE OCEAN FLOOR The breakthrough that made sense of continental drift was the work of the American geologist Harry Hammond Hess, who was born in New York in 1906. He studied geology at Yale, and worked as a geologist in what is now Zambia. In 1934 he joined the faculty at Princeton, becoming professor of geology in 1950. He later advised NASA, and helped to plan the first Moon landing.

In 1956, the Texan geologist William Maurice Ewing showed that a 55,000-kilometre/35,000-mile long mountain ridge wound its way right round the world, in the middle of the ocean floor. In 1957, he established that this was split throughout its entire length by a massive rift. Hess combined Ewing's findings with his own discovery that the rocks of the ocean floor were much younger than the rocks of the continental crust. This enabled him to come up with an explanation of the origin of the ocean bed, which he advanced in a paper entitled *History of Ocean Basins* in 1962. His theory was that molten rock was being forced up at the mid-ocean rifts, and that the pressure of this newly formed rock was forcing the older rocks of the ocean bed apart. It was this process – which he called *sea-floor spreading* – that was driving the continents apart.

In 1963, two British geologists, Fred Vine and Drummond Matthews, published the results of their investigations into the magnetic orientation of rocks in the ocean bed. It was already known that the direction of the Earth's magnetic poles was subject to sudden reversals from time to time; and Vine and Matthews established that the ocean floor displayed alternating stripes of 'fossilized' polarity that were symmetrical on either side of the mid-ocean rifts. These could only have originated when the various segments of ocean floor were in a molten condition. By the end of the 1960s, a new science – *plate tectonics* – had been created, and continental drift, which had seemed an outlandish proposition, was accepted as a fact. We know now that the continents are not 'drifting', but being carried on 'plates', which move in response to the pressure generated by this newly formed rock. In the Atlantic, this has the effect of pushing the continent-carrying plates apart. The movement is only a few centimetres a year. But over long periods the plates travel huge distances. Where they collide, one of two things can happen. In some places, such as the West Coast of North America, the pressure of one plate against another creates stresses that cause earthquakes. In other places, such as where the plate carrying the Indian subcontinent presses against the plate carrying Asia, the Earth's surface buckles, and mountain ranges are formed.

TECTONIC PLATES This term first appeared in a journal article in 1968, and the term rapidly replaced 'continental drift', which was not a scientific concept, just a crude description of a visible phenomenon. Continents are defined by their ocean boundaries, and there is little correlation between the outlines of the continents and the edges of the plates that carry them. New Zealand is carried on the same plate as the bed of the Indian Ocean, and Iceland stands astride the point of contact of two plates. There are many more plates than there are continents: about a dozen large ones, and 20 small ones. The annual movement of the individual plates is measured in centimetres; but, over millions of years, centimetres add up. Parts of the Earth's crust that once sweltered under a tropical sun are now buried under polar ice. And when immense masses of rock collide – even at a snail's pace – the pressure, and the friction, can have dramatic consequences. The earthquakes and volcanoes that are a feature of the regions where plates meet are visible signs of the forces at work.

EARTHQUAKE ZONES Some earthquakes are associated with volcanic eruptions, but most are the result of the release of stresses that have built up in the Earth's crust. These stresses are generated by tectonic activity in two types of situation: where the continent-carrying plates are in collision, and where new material is being forced up through the ocean bed. About 80 per cent of all destructive earthquakes occur in a ring around the Pacific Ocean, and most of the others occur in a band that stretches along the Mediterranean, and on through the Middle East and South Asia. The epicentre of an earthquake may be as much as 600 kilometres/400 miles below ground, but the ones that do the most damage occur in the top 50 kilometres/ 30 miles of the Earth's crust.

THE SEISMOGRAPH The strength and duration of an earth tremor is recorded by a device called a seismograph (or seismometer). The credit for the invention of the seismograph is usually given to the Italian physicist, Luigi Palmeri, who constructed one in 1855, based on the movements of mercury in a sealed tube. But this claim is questionable. Palmieri's instrument could not distinguish between local disturbances, such as a passing heavy cart, and a distant earth tremor. The first seismograph really deserving of the name – and the forerunner of all present-day instruments – was the one constructed by the English geologist John Milne in 1880.

THE RICHTER SCALE Earthquakes are measured on the Richter scale. This was devised in 1935 by an American seismologist, Charles F. Richter. The problem with measuring earth tremors is that the largest can be 500 million times more powerful than the smallest that can be detected. Richter got round this by devising a logarithmic scale of 0 to 10, wherein each step of the scale represents a 10-fold increase in magnitude. The scale measures the size of the seismograph traces, not the energy of the earthquake. The seismic wave generated by a tremor of force 6 is not 20 per cent higher than a wave of force 5, but 10 times higher. The energy involved may be 50 times greater. A 'major' earthquake – 7 on the Richter scale –

occurs somewhere in the world about once a month. An 'great' earthquake of force 8.5 – something like 200 times more powerful – happens about once every 10 years. The two most powerful earthquakes recorded since modern measurement began were those at Sanriku, in Japan, in 1933, which registered 8.9, and the earthquake experienced in Southern Chile in May 1960, which was originally estimated at 8.6, but was later revised to 9.5.

Force and damage are not the same thing. A hulk full of high explosive, detonated on the high seas, would cause a splash but little damage. In a crowded port, it would constitute a disaster. Similarly, a force 7 tremor far underground in the Arctic may be a polar bear's disturbed night. Close to the surface in central Tokyo, it would be a human catastrophe. The most destructive earthquake on record occurred in China's Shanshi province in 1556. It spread destruction over an area 500 kilometres/300 miles in radius, and caused the death of an estimated 800,000 people. It was by a long way the greatest natural disaster in history.

THE TSUNAMI Among the most spectacular products of earth movements are the ocean waves called tsunami (meaning 'big waves' in Japanese). Tsunami may travel thousands of miles from the site of the earthquake that causes them. Some are the product of earthquakes on land. Others originate in the depth of the ocean. They travel at great speed – anything from 150 to 800 kilometres an hour/100 to 500 miles an hour. At sea, they take the form of a succession of swells, 1 metre/3 feet high at most, which a ship may hardly notice. But when they reach a shore, they slow down, and pile up on top of one another. If they are forced into river estuaries, they can turn into walls of water, 15 to 30 metres/50 or 100 feet high, capable of sweeping away towns, and tossing ships inland. Distance does little to lessen their force: an earthquake on the Aleutian Islands, off Alaska, in May 1946, hurled waves 15 metres/50 feet high onto the beaches of Hawaii, 3,000 kilometres/2,000 miles away.

The most terrifying example in recent times of the power of tsunami was the earthquake off the island of Sumatra in Indonesia in December 2004. This occurred at the boundary where the plate carrying India is forced under the much smaller Burma plate. The average annual movement of the Indian plate is only of the order of 5 centimetres/2 inches; but there had been virtually no movement in the region of Sumatra for 150 years, during which time enormous pressures had built up. On 26 December, they were released when a slippage occurred about 10 kilometres/6 miles below the ocean bed, registering 9.0 on the Richter scale. The Burma plate sprang up by about 1.5 metres/5 feet in a few seconds, displacing enormous volumes of water, and sending a series of tsunami 1 metre/3 feet-high hurtling across the Indian Ocean. Six thousand kilometres/four thousand miles away, in Somalia, they claimed the lives of nearly 200 people. On the coast of Sumatra the devastation was total. The city of Banda Aceh, home to 400,000 people, was utterly destroyed within minutes of the earthquake, and 90,000 of its inhabitants died. The immediate loss of life around the Indian Ocean was more than 200,000, making it one of the most destructive natural disasters of the past 500 years.

SOME NOTABLE EARTHQUAKES

Year	Location	Force	Estimated deaths
856	Damghan, Iran	–	200,000
1556	Shansi, China	–	800,000
1737	Calcutta, India	–	300,000
1755	Lisbon, Portugal	*8.7	70,000
1812	New Madrid, USA	*7.9	Very few
1906	San Francisco, USA	*7.7	3,000
1920	Gansu, China	*8.6	200,000
1932	Gansu, China	7.6	70,000
1933	Sanriku, Japan	8.9	3,000
1960	Southern Chile	9.5	6,000
1970	Northern Peru	7.7	85,000
1976	Tangshan, China	8.5	250,000**
1988	North-west Armenia	6.8	55,000
1990	Northern Iran	7.7	35,000
1995	Kobe, Japan	6.9	5,000
1999	North-west Turkey	8.2	20,000

* Estimated

** True figure probably much greaer

VOLCANIC ERUPTIONS Two zones of earthquake activity – around the Pacific, and from Indonesia west to the Mediterranean – are the home of most of the world's active volcanoes. This is no coincidence. The forces that create the earthquakes are the same forces that generate the heat that issues in volcanic eruptions. As with major earthquakes, by far the greatest number – two-thirds – of all volcanic eruptions occur in a circle around the Pacific, earning it the evocative name of the 'Ring of Fire'.

Until quite recently, it was thought that the lava that spewed forth from volcanoes was molten rock from deep in the Earth's core, which had somehow found its way out through weaknesses in the planet's crust. That theory has now been abandoned. A version of it is still used to explain a small number of isolated volcanoes, such as those in Hawaii, where a particularly thin layer of crust overlies a 'hot spot' in the Earth's mantle. But the heat energy that fires the volcanoes of the Pacific Rim is generated where the rocks of one plate meet the rocks of another, and are forced below it, melting under the enormous friction to which they are subjected.

There are about 500 active, or potentially active, volcanoes in the world. But 'about' is the operative word. It is difficult to be sure whether a volcano that has shown no signs of activity for hundreds of years is extinct or just dormant. It used to be supposed that there were no potentially active major volcanoes in the continental United States. But on 18 May 1980, a sleeping giant named Mount St Helens in Washington State, which had not twitched for 126 years, exploded in an eruption that took 120 metres/400 feet off its height, and blew away an entire side of the mountain.

If that seems impressive, it was nothing compared to what happened in the Sunda Strait in Indonesia, on 27 August 1883, when the island of Krakatoa vanished in an explosion that blew 20 cubic kilometres/5 cubic miles of rock and ash into the sky. It did so with a bang that was heard in Australia, 3,000 kilometres/2,000 miles to the southeast, and on Rodriguez Island, 5,000 kilometres/3,000 miles to the southwest; and it created tsunami 40 metres/120 feet high on the coasts of the neighbouring islands of Java and Sumatra. The coastal towns of the two islands were then much less densely populated than they were in 2004; nevertheless, 36,000 people were swept to their deaths.

But even Krakatoa could not compare with the eruption 1,200 kilometres/700 miles to the east, on the island of Sumbawa, on 11 April 1815. On that day, a volcano called Tambora exploded in the most massive eruption of the last 20,000 years. Around 120 cubic kilometres/30 cubic miles of pumice and ash were ejected in the explosion, and the entire mountain, which had stood 4,000 metres/13,000 feet above the sea, dropped into the chamber that the ejected material had left behind. Some 10,000 people were killed in the eruption, and another 90,000 succumbed to starvation and disease in the resulting dislocation. A cloud of dust and ash spread around the Earth, obscuring the sunlight. In 1816, crops failed in country after country; for years afterwards it was remembered in the northeastern United States as 'Eighteen Hundred and Froze to Death'.

COAL AND OIL One legacy of past movements of the Earth's crust are the coalfields scattered around the globe. Coal is not a mineral. It is the fossilized remains of plants that flourished around 200 to 300 million years ago, in the periods known to geologists as the Permian and the Carboniferous.

The trees whose trunks and leaves have left their impressions in the coal grew in a warm, humid climate. When they died, they rotted down in boggy soils, creating vast deposits of peat. Earth movements caused these deposits to sink beneath the sea. Over millions of years, they were buried by sediments washed down from the land; as these sediments accumulated, the pressure of their weight transformed the organic material into coal. Finally, an uplifting of the ocean bed brought the coal-bearing strata back above sea level again. A similar process gave us the oilfields of Texas, Central Asia, and the Persian Gulf, but in these cases the organic source was not plants that had lived on land, but plants and animals that had died in the sea.

THE EARTH'S TOP LAYER Most of the material that makes up the Earth's crust is composed of a small number of elements, of which oxygen and silicon are by far and away the most abundant:

Element	% (by mass)
Oxygen	46.1
Silicon	28.2
Iron	5.6
Calcium	4.2
Sodium	2.4
Magnesium	2.3
Potassium	2.1
Aluminium	0.8
Titanium	0.6
Other elements	7.7
	100.0

The high silicon content in the surface rocks is accounted for by the prevalence of the mineral *silica* (SO_2), in the form of quartz, sand, etc. It is believed that iron accounts for a significant proportion of the mass at lower depths.

MEASURING HARDNESS The hardness of rocks is measured by Mohs' scale. It was devised by the German mineralogist Friedrich Mohs (1773–1839), who published it in 1812. It is based on a reference list of ten familiar minerals, as follows:

1. Talc	6. Feldspar
2. Gypsum	7. Quartz
3. Calcite	8. Topaz
4. Fluorite	9. Corundum
5. Apatite	10. Diamond

The grade of a mineral not on Mohs' list is determined by its ability to scratch minerals that are – and to scratch minerals of equal hardness. A mineral midway in hardness between feldspar and quartz would be graded 6.5. A fingernail is 2.5, midway between gypsum and calcite.

The hardness of some everyday materials depends very much on their composition. Harder varieties of glass, which may range up to 6.5, easily scratch mild steel, which is seldom harder than 5.0. Most penknives, on the other hand, are made of hardened steel graded around 6.5, and easily scratch the softer kinds of glass.

EXPLORING THE EARTH'S INTERIOR Geologists have found many uses for seismometers. One of these is the investigation of the Earth's interior. This is studied by creating what are in effect tiny earthquakes, and measuring the path taken by the resulting waves through the Earth's rocks, and time they take to complete their journeys. These investigations are supplemented by studies of the waves created by actual earthquakes.

As a result of these studies, we now have a fair idea – though no more than that – of the Earth's interior, and the variation in its composition between the surface and the core. But 'fair' is the operative word.

Most geologists would accept the following as a rough description of the Earth's interior:

	Thickness (km)	(miles)	State	Density (g/cm^3)	Temperature (° C)	(° F)
Crust	0–15	0–25	Solid	2.8	up to: 550	1,000
Upper mantle	c.650	c.400	Molten	4.3	up to: 800	1,500
Inner mantle	c.2,100	c.1,300	Solid	5.5	up to: 2,500	4,500
Outer core	c.2,100	c.1,300	Molten	10.0	up to: 3,000	5,500
Inner core	c.1,500	c.900	Solid	13.5	up to: 2,750	5,000

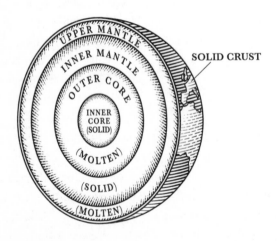

Figure 17. The Interior of the Earth

WATER AND SALT   Compared with other planets, the Earth is a wet place. The oceans cover more than 70 per cent of its surface. The Sun's heat causes this water to evaporate. Some of it returns as rain over the ocean. Winds carry water vapour over the land, where it condenses and falls as rain or snow. Much of this is evaporated by the Sun, but some percolates into the soil. Some of the water in the soil is taken up by plants, and returned to the atmosphere through their leaves. Some runs off into rivers and back to the sea. And some trickles down through the soil and into the underlying rocks, where it forms ground water. The water that runs off the land carries away quantities of rock and soil; over thousands of years this changes the appearance of the land.

About three per cent of the weight of sea water is accounted for by solids, of which three-quarters is common salt. The salt content of the ocean is being continually added to by run-off from the land, but it is not increasing. Yet salt cannot leave the sea by evaporation, because the water vapour leaves it behind. So for many years it was a mystery as to why the sea should have so little salt in it – given the rate of run-off – and par-ticularly if the world was supposed to be hundreds of millions of years old. The riddle wasn't solved until the 1970s, when the deep ocean vents were discovered. Sea water is swallowed up by these fissures in the ocean bed; when it re-emerges, millions of years later, as steam from volcanoes, the salt has been filtered out of it on its passage through the rocks.

It seems that the overall composition of sea water has not changed to any marked extent for the past 100 million years. This doesn't just apply to its salt content; it also applies to other dissolved substances, and to the insoluble products of land erosion, which don't stay in suspension indefinitely, but gradually settle on the ocean floor. However, industrial products are a different matter. As humanity increases its output of plastics and fertilizers, waste metals and medicines, the sea's burden of harmful chemicals increases, and the fish and corals that live in shallow seas shrivel before the onslaught.

Three-quarters of the all the fresh water on land is not in the form of water, but ice. And two-thirds of all the Earth's fresh water sits virtually motionless on the Antarctic ice cap. The ice sheets of Antarctica are, on average, 2,000 metres/7,000 feet thick. And they are millions of years old. Contrary to what one might think, little snow falls on Antarctica. It is a vast desert, like the Sahara. However, unlike the Sahara, it has been a desert for a very long time.

CORAL REEFS An intriguing feature of the oceans is the coral reefs found in tropical seas. They are composed of the skeletons of millions of creatures called polyps. Coral polyps form skeletons around themselves; and also buds, which remain attached to the parent skeleton. These buds in turn form skeletons, and more buds, and so the process continues. In time a great mass of skeletons is formed, and this forms the reef.

Polyps live only in warm, shallow seas, down to about 90 metres/300 feet. Yet some coral reefs extend to depths of 1,500 metres/5,000 feet. This used to be a complete mystery. There is no single accepted theory as to the causes of reef formation. Some reefs seem to owe their existence to fluctuations of sea level associated with past ice ages. But the theory that provides the best explanation for the majority of reefs was that advanced in 1842 by Charles Darwin, in his book *The*

Structure and Distribution of Coral Reefs. He suggested that reefs are formed where the sea bed is subjected to gradual subsidence, as a result of volcanic action. As the reefs sink, successive generations of polyps make their home on the remains of their ancestors.

TRIANGULATION AND PARALLAX Big questions can sometimes be answered with simple maths. Suppose one wishes to know how far it is to a distant mountain (P). The easiest way to find out is to use triangulation. A 'baseline' is laid out between two points, A and B, whose distance apart is known. From A, the angle between P and B is measured. Then from B, the angle between P and A is measured. These two angles enable one to construct the triangle ABP, and to calculate the length of sides AP and BP. This is triangulation, the basic tool of mapmaking. The angle APB is the *parallax* of the peak.

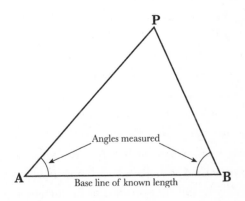

Figure 18. Basic Triangulation
Finding the distance to a mountain peak from a measured baseline.

The usefulness of parallax is not confined to the Earth's surface. In 150 BC, the Greek astronomer Hipparchus used it to calculate the distance to the Moon, from the size of the Earth's shadow during a lunar eclipse. His estimate was that the distance to the Moon was nearly 59 times the Earth's equatorial radius. The Earth's radius is 6,400 kilometres/4,000 miles, which would imply a distance to the Moon of around 378,000 kilometres/235,000 miles. This was within 2 per cent of the correct value, which is 384,000 kilometres/239,000 miles.

MEASURING ASTRONOMICAL DISTANCES Using parallax to measure the relative distance of the Sun and Moon was an impressive achievement 2,000 years ago. It was not until 1672 that the next step – measuring the distance to the Sun – was taken. Once this was achieved, it was natural that astronomers should try to repeat the trick with a star. Stars are much farther away than the Sun, so a longer baseline is required: longer than any line that can be laid out on the Earth's surface. And even if it turned out to be possible to measure the distance to a star, it would have to be one close enough to display a measurable parallax.

Fortunately, a bigger baseline is available: the width of the Earth's orbit. The Earth is 150 million kilometres/93 million miles from the Sun. If observations are made, six months apart, on opposite sides of the Earth's orbit, this yields a baseline 300 million kilometres/186 million miles long. Even so, the angle to be measured is still extremely small. But in 1784 the feat was accomplished, by a German astronomer named Friedrich Wilhelm Bessel.

Bessel was born in Prussia in 1784. He was an accountant who taught himself mathematics and astronomy. When he was 20, he recalculated the orbit of Halley's comet, an achievement that gained him a post in an observatory. In 1810, he was appointed by Frederick Wilhelm III to create a new observatory at Königsberg in Prussia, where he continued as director for the next 30 years.

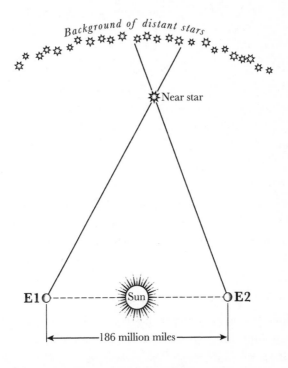

Figure 19. Using Parallax to Measure the Distance to a Star
When viewed from positions E_1 and E_2 on opposite sides of the Earth's orbit, the nearer star appears in a different place against the background of more distant stars.

Some stars have a discernible *proper motion* – a slight change of position, over a period of years, in relation to the stars around them – betraying the fact that they are nearer to the Earth than these other stars. The best hope of finding a star with a measurable parallax was to find a star with a sizable proper motion. In 1838, Bessel decided to concentrate his attention on a star – 61 Cygni – in the constellation of the Swan. It was a very faint star, but it had an exceptionally large proper motion. Using an instrument of his own design, he succeeded in measuring its parallax, relative to two neighbouring stars with no discernible proper motion at all. The angle involved was extraordinarily small: $\frac{1}{10,000}$th of a degree – the diameter of a small coin seen from a distance of 5 kilometres/ 3 miles. It was an angle that implied a distance from Earth of some 56 million billion kilometres/35 million billion miles – or 11 light years.

During the next 70 years, more stars with measurable parallaxes were discovered. By 1900, the distances of 75 had been determined. Only a few were as close as 61 Cygni. The nearest of all was Alpha Centauri, just four light years distant. This was twice as far as Isaac Newton had placed the boundaries of space.

THE DOPPLER EFFECT   Anyone who has stood on a station platform while a whistling locomotive passes by will know that the sound of its whistle drops in pitch after it has passed. This phenomenon – the *Doppler effect* – is named after the man who first analysed it: an Austrian physicist named Christian Doppler. He was born in Salzburg in 1803, and he was professor of experimental physics at the Physical Institute in Vienna.

Doppler speculated that the frequency of the sound waves from a moving source would depend upon the motion of that source, so that more waves would reach the ear in a given time from an approaching sound source than from a receding one. If so, the higher frequency would by definition be registered by the ear as a more highly pitched sound. In 1842, he put forward a mathematical theory relating the speed of a moving sound source to the frequency of the sound perceived. This was verified two years later by a picturesque experiment, in which a locomotive pulled a wagon carrying a group of trumpeters backwards and forwards at varying speeds over a period of two days, while a party of musicians noted down the sounds they heard.

Doppler next suggested that *light* emitted by a moving source would be similarly affected, and that the reduced frequency of the light waves received from a receding source would result in a reddening of its light. In 1868 the British astronomer William Huggins provided confirmation of this proposition, when he detected a 'red shift' in the spectrum of the star Sirius.

The Doppler effect in the case of light is technically a *Doppler–Fizeau effect*. This commemorates the French physicist Hippolyte Fizeau, who pointed out in 1848 that the effect would manifest itself as a shift in the lines in a star's spectrum. In the case of a receding star, the lower frequency causes the lines to be shifted towards the red end of the spectrum – a red shift. The lines in the spectrum of an approaching star display a blue shift. The shift is proportional to the speed at which the object is approaching or receding; and it can be used to measure that speed. It was this that enabled Huggins to put a value on the speed at which Sirius was moving away from the Earth

– or rather, the speed at which Sirius and the Earth were moving apart.

HENRIETTA LEAVITT AND THE HARVARD COMPUTERS

In the three centuries after Copernicus, a succession of discoveries revealed the insignificance of the Sun, and the entire solar system, when set against the crowd of stars that made up the Milky Way. As the twentieth century dawned, astronomers could have been forgiven for thinking that they had at last got a grasp of the scale of the universe. But the biggest shocks were still to come.

In 1876, a 30-year-old professor of physics at Massachusetts Institute of Technology named Edward Charles Pickering was appointed director of the Harvard University Observatory. In 1891, with the aid of his younger brother, William, he established an outstation of the observatory at Arequipa in southern Peru. Pickering was an enthusiast for the new photographic approach to the mapping of the heavens. Using photometers of his own invention, he supervised a survey that succeeded in mapping the positions of 45,000 stars. The observations were processed at Harvard, by a team of women cataloguers who were known as the 'Harvard Computers'. Among their number was a young woman named Henrietta Leavitt, whose love of astronomy was such that she joined the team as an unpaid volunteer. Her outstanding ability soon led to her being appointed head of the observatory's stellar photometry unit.

Leavitt's special interest was the study of Cepheid variables. This class of star was named after Delta Cephei, a star described by John Goodricke in the 1780s. Cepheid variables display a pattern of behaviour characterized by a sudden increase in luminosity followed by a gradual dimming, repeated in a regular cycle. The variation is caused by a periodic pulsation in the light emitted by the star.

In 1902, Leavitt began a study of Cepheid variables in the Smaller Magellanic Cloud, a cloud of stars visible from the observatory in Arequipa. Cepheid variables have a 'period', or cycle of fluctuation, that can be anything from a day to three months in length; and Leavitt noticed a correlation between the length of a Cepheid's period and its brightness. In 1912, she published a graph showing that, if one plotted the logarithm of the length of a Cepheid's period against its apparent luminosity, the result was a straight line. This was a result of great significance, since it provided a new way of ascertaining the distance of far-off stars.

THE STARS' BRIGHTNESS AND DISTANCE The chain of reasoning that turned Leavitt's discovery into a measuring tape for stellar distances was as follows:

1. The apparent magnitude (that is, the perceived, as opposed to the real, brightness) of a star depends on its distance from Earth.

2. The Smaller Magellanic Cloud was so far away that all the stars in it were effectively the same distance from the Earth.

3. Therefore the correlation between the period and the apparent brightness of Cepheid variables in the cloud must reflect an identical correlation between their periods and their real brightness.

4. If the distance of one or two nearby Cepheid variables could be ascertained by another method, such as parallax,

their real brightness could be calculated from their apparent magnitude.

5. A comparison of the periods of these 'reference' stars with their absolute (real) brightness would make it possible to construct a scale from which one could derive the real brightness of any Cepheid variable, no matter how distant, from the length of its period.

6. The comparison of a Cepheid's real brightness with its apparent magnitude would then indicate its distance from the Earth.

EINSTEIN'S IMPORTANT PAPERS In 1902, the year in which Leavitt started her researches into Cepheid variables, a 23-year-old clerk was just settling in to his new job as a junior examiner in the Patent Office in the Swiss city of Zurich.

The name of the young clerk was Albert Einstein. He had been born in Ulm, in Germany, in 1879, but went to school in Munich, where his father owned a small factory. He was bored at school; but an uncle aroused his interest in mathematics, which would remain a passion for the rest of his life. When the family later moved to Italy, Einstein continued his schooling in Aarau, Switzerland. When he was 17, he entered the Federal Institute of Technology in Zurich, to train as a science teacher. He did reasonably well in his examinations, but his attitude did not endear him to his tutors, one of whom went so far as to say, 'You will never amount to anything, Einstein.'

Einstein was a pacifist; and had moved to Switzerland to avoid military service. On his graduation in 1900 he had become a Swiss citizen. He had been unable to find an academic post, and had been lucky to get the Patent Office job through the influence of a friend's father. The job had its advantages. It was not very demanding, and it gave him lots of time to think about the mathematical problems that interested him. And just as Newton, at the same age, was already doing highly original work in mathematics, so Einstein was already working, armed only with pencil and paper, on a range of challenging problems in theoretical physics.

In 1905, when he was 26, he burst upon the world of science with a set of papers of the highest originality, which were published in the German *Yearbook of Physics*. One dealt with the *photoelectric effect*, in which light falling on certain metals stimulates the emission of electrons. This made a major contribution to the development of the photon theory of light. It was a piece of work that would lead to his being awarded the Nobel Prize for physics in 1926; yet it was not the most important of the papers he published in that year.

His second paper presented an analysis of *Brownian motion,* the strange, jerky movement made by tiny particles of substances, such as pollen, when they are suspended in water. This paper presented an equation that explained, in precise form, the slow drift of such particles from their original starting point, as they reacted to collisions with the molecules of the liquid in which they were suspended. It would later play an important part in the development of twentieth-century atomic theory.

Another of these papers introduced what Einstein called his *special theory of relativity,* questioning some ideas of Newton's that had remained unchallenged for 250 years. The special theory did not deal with gravitation. Einstein would go on to challenge aspects of Newton's gravitational theory as well in his 1916 *general theory of relativity.* The special theory was concerned with concepts of motion and of time. A central role in the theory was allotted to the behaviour of light, and to the bizarre consequences for objects that moved at speeds approaching the speed of light. According to this theory, the speed of light had a special significance. In the first place, it was the same for all observers, no matter what their own motion. Secondly, it was not additive. Unlike, for example, a cannonball shot from a ship, which has a speed that is a combination of its own muzzle velocity and the motion of the ship; light does not partake of the motion of its source, whatever the motion of that source.

It was possible to design experiments to test these propositions; and in the years following the publication of the special theory, a number were carried out with great care. The result was a resounding confirmation of Einstein's propositions; the special theory is now a cornerstone of modern physics.

THE FAMOUS EQUATION Einstein's special theory promoted one equation that has become famous: $e = mc^2$. The significance of it was demonstrated by the detonation of the atomic weapons that destroyed the Japanese cities of Hiroshima and Nagasaki 40 years after Einstein's paper was published. Unlike most of the special theory, the message embodied in these few symbols is capable of being understood by anyone with even a modest scientific education. The symbols stand for: e (energy); m (mass); c (the speed of light).

According to Einstein's theory, which had its devastating confirmation on those August days in 1945, atoms contain immense stores of energy locked up in the tight embrace of their particles. If these particles are forcibly rearranged, as they are in the detonation of an atomic bomb, a small amount of mass is lost, but an enormous amount of energy is released. In his equation, c – the speed of light in a vacuum – is 300,000 km per second/186,300 miles per second. Therefore, c^2 is either 300,000 × 300,000, or 186,300 × 186,300, depending on whether energy, mass, and speed are expressed in International (SI) Units, or in British (fps) units. These are very large numbers indeed, and they indicate the phenomenal amount of energy released when a tiny amount of mass is lost.

CLASSIFYING THE STARS In the early twentieth century, research done at Harvard on the spectra of stars came to the attention of Henry Norris Russell, director of the Princeton Observatory.

Russell was interested in the life cycles of stars; it occurred to him that the classification of spectra made by the Harvard researchers might provide a key to the history of individual stars. In 1914, he published a graph in which he plotted the spectral class (that is, the colour) of a number of stars against their real luminosity (their true, as opposed to their apparent brightness). The graph displayed a striking pattern. Blue light is an indicator of higher temperature than red light. On the graph, most stars fell within a band stretching from the top left-hand corner to the bottom right-hand corner: that is, from hot and bright to cool and faint. The implication was that there was a correlation between a star's temperature and its size.

A Danish astronomer, Ejnar Hertzsprung, had made a similar suggestion some years earlier; and in recognition of their joint contribution, the graph became known as the Hertzsprung–Russell diagram. The band within which the majority of stars lay became known as the 'main sequence'; and stars that lie within it – such as the Sun – are still referred to as main sequence stars. There are two groups of stars that do not lie within the main sequence: red giants and white dwarfs. Red giants, such as Betelgeuse, the brilliant star in the constellation of Orion, are huge globes of comparatively cool gases that owe their brightness to their size. Betelgeuse is so big that, if the Sun were at its centre, the entire orbit of the Earth would lie within the star. White dwarfs, which are quite common, are

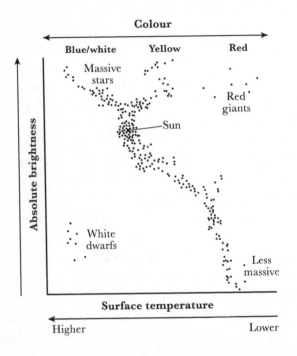

Figure 20. The Hertzsprung–Russell Diagram, Relating Luminosity to Surface Temperature
The Sun is an average sort of 'main sequence' star.

small and hot. They are also solid. A white dwarf smaller than the Earth may have a mass greater than that of the Sun.

In the early years after the publication of the diagram, there was a temptation to regard the main sequence as the path that the most stars followed as they aged. But as understanding of the internal processes of stars grew, it became clear that the truth was more complicated. It is no longer seen as a picture of the life cycle of a 'typical' star; but it still provides a useful framework for the discussion of stellar evolution.

EINSTEIN'S GENERAL THEORY   It had taken Einstein four years after the publication of his special theory to obtain an academic post; it was not until 1909 that he was appointed to a modestly paid professorship in the University of Zurich. But his international reputation continued to grow, and in 1913 he was offered a specially created post in the Kaiser Wilhelm Institute in Berlin, where he had the benefit of stimulating contact with several of the world's most eminent physicists. When the First World War broke out, he was protected by his Swiss citizenship from having to serve in the German forces, and was able to continue his researches. In 1916, he published another great paper, which is usually referred to by the name of the theory it expounded: *the general theory of relativity.* The word 'general' derived from his aim to extend his special theory to include gravitation. His original theory had been particularly concerned with phenomena at the atomic and subatomic level. He now extended its scope so that it applied to the universe as a whole. In doing so, he came up against certain problems with Newton's laws.

It is sometimes said that, in formulating his general theory, Einstein 'overturned' Newton's laws. This is an exaggeration. For everyday purposes, and for most problems on the scale of the solar system, the difference between the answers given by the two theories is negligible. But at extremely high speeds, and in the presence of extremely large gravitational forces, Einstein's theory wins hands down.

One of the differences between the two theories lies in the explanation of the nature of gravitational attraction. For Newton, gravity was a force emanating from massive objects, and transmitted by a medium which he called the 'ether', which pervaded the whole of space. In Einstein's view, the ether did not exist: it was a fiction invented to make 'action at a distance' acceptable. In his analysis, gravitation was not a force residing in objects that possessed mass, it was the consequence of the bending of space in the neighbourhood of mass. In the case of a body like the Sun, the image often employed is of a heavy ball lying in the middle of a suspended sheet of rubber. The ball will create a hollow in the sheet, and if a smaller ball is rolled towards the larger one, it will be deflected into a curve as it rolls past. In Einstein's theory, the space around a large body like the Sun is distorted by the Sun's mass, and the elliptical orbit of a planet is the consequence of this distortion.

There are several ways in which the relative merits of Newton's and Einstein's theories can be tested. One of the most famous was an experiment conducted during a total eclipse of the Sun to ascertain whether starlight passing close to the Sun was deflected by the Sun's gravity. It was found that the light was deflected by precisely the amount that Einstein had said it would be.

In 1930, Einstein travelled to America to lecture at the California Institute of Technology. He was still there when Hitler came to power in Germany. As a Jew, it would have been folly for him to return to Germany. He took up residence at Princeton Institute of Advanced Studies in New Jersey, and in 1940 became an American citizen. His later years were devoted to an unsuccessful search for a 'unified theory' that would explain both gravitation and electromagnetism. After the Second World War he was active in the campaign to restrict the spread of the nuclear weapons that provided such devastating proof of the validity of his most famous equation.

HUBBLE AT MOUNT WILSON In the years following the publication of Leavitt's scale, astronomers established the distance of several nearby Cepheid variables. Using her formula, they were then able to ascertain the distance to stars in the farthest reaches of the Galaxy, and to determine the size and shape of the Galaxy itself. One astronomer succeeded in applying Leavitt's measuring stick to stars beyond the Galaxy, with consequences that changed humanity's view of the universe. His name was Edwin Hubble.

Hubble, who was born in Marshfield, Missouri in 1889, was the son of a lawyer, and was himself intended for a legal career. He won a Rhodes Scholarship to Oxford, and studied law there. But he became interested in astronomy and, after finishing his degree, he joined the staff of the Yerkes Observatory, near Chicago, where he worked from 1914 to 1917. On his return from war service in 1919, he was offered a post in the new Mount Wilson observatory, where he had access to the

254-centimetre/100-inch reflecting telescope, then the most powerful instrument of its kind in the world.

Early in his career at Mount Wilson, his attention was attracted to nebulae, the fog-like objects found in every corner of the heavens, most of which had defied previous efforts to resolve them into greater detail. At this time, the shape and size of the Galaxy were reasonably well understood, but it was not known what – if anything – lay beyond its boundaries. The word *galaxy* was of Greek origin, and meant, literally, the Milky Way. In the early twentieth century the two terms were still used interchangeably, and the Galaxy was treated as being synonymous with the visible universe.

It was clear that some nebulae were located within the Galaxy, and that they were basically clouds of gas illuminated by stars within them. In 1924 Hubble succeeded in making out individual stars in the Andromeda nebula, some of which he identified as Cepheid variables. Using Leavitt's period–luminosity law, he was able to arrive at an estimate of their distance, which he calculated at 800,000 light years, 8 times farther off than the most distant stars previously known. (This would later turn out to be a serious underestimate.) During the next few years, he repeated this success with one nebula after another, until it became clear that the Galaxy was just one of a host of 'island universes', each containing vast numbers of stars.

THE EXPANDING UNIVERSE Had this transformation of our image of the universe been Hubble's only achievement, it would have been enough to earn him lasting fame; but there was more to come. In the half-century since Huggins had recorded the red shift in the spectrum of Sirius, a number of red and blue shifts had been measured in the spectra of objects within the Galaxy. But an astronomer at the Lowell Observatory named Vesto Slipher had discovered that virtually all the extra-galactic nebulae – the 'island universes' beyond our own Galaxy – whose spectra he had been able to analyse, displayed significant red shifts. It was a result that no one could explain.

In 1929, Hubble published an analysis of the radial velocities of nebulae whose distances he had calculated. These were their velocities in the line of sight from the Earth, as deduced from the red shift in their spectra. If his earlier discovery of a host of extragalactic nebulae had caused astronomers to revise their image of the cosmos, his latest idea made their heads spin. What he had established was that, although a few extragalactic nebulae had spectra that indicated they were moving *towards* the Earth, the great majority showed red shifts that could only be explained on the assumption that they were moving *away*. Even more startling was his discovery of a direct relationship between a nebula's distance and its speed of recession.

Hubble concluded that the only explanation consistent with the red shifts recorded was that, apart from a 'local group' of nearby galaxies, the extragalactic nebulae were all receding from the Galaxy; and that the farther off they were, the faster they were moving away. This only made sense on one of two suppositions: either there was something unique about the Galaxy and its position in space; or the universe itself, including the space between the galaxies, was expanding. Hubble was in no doubt as to the answer: the evidence pointed unequivocally to the conclusion that we live in an expanding universe.

A NEW VIEW OF THE UNIVERSE Hubble's suggestion that the universe itself was expanding called for such a reorientation of cosmology that some astronomers were initially reluctant to accept it. But by the end of the 1930s, Hubble's interpretation was accepted by all those best qualified to judge it.

Any analogy is bound to distort the reality of such a profound process, which can only really be understood mathematically. But for the non-mathematician, one way of understanding it is to think of ants scurrying about on the surface of a large balloon that is being inflated. The ants have their individual movements, and at any given moment particular ants may be seen to approach one another. But, on average, the ants will be seen to be moving farther apart; and the ants that are farthest apart will, on average, be moving apart at the fastest rate. In this analogy, the ants are the individual nebulae, or star clusters, and the balloon is the universe. Individual nebulae have their own movements (called *peculiar motions*), but these are superimposed on a phenomenon that affects them all: the expansion of the universe.

OUR LOCAL GROUP OF GALAXIES Not all the extragalactic nebulae are moving away from us. There is a group of galaxies called the *local group* that revolve around a common centre of gravity they share with our Milky Way Galaxy. This group includes the Great Nebula in Andromeda (which is much bigger than our own Galaxy), the Clouds of Magellan, and about two dozen mostly smaller ones. Apart from these, all other galaxies have a movement away from us that reflects the expansion of the universe.

LEMAÎTRE'S 'COSMIC EGG'   The acceptance of the reality of an expanding universe led naturally to the question of how long the expansion had been going on. Logically, it had either started at a particular moment in the past, or it had been going on for ever. Either option raised further questions of a kind that physicists had never faced before.

One of the first to throw his hat into the ring was the astronomer, Georges Édouard Lemaître. Lemaître was born in the Belgian town of Charleroi in 1894. He was a civil engineer who became interested in physics and mathematics while serving as an artillery officer in the First World War. After the war, he took a Ph.D. at his old university of Louvain, and was then ordained as a priest. He studied astrophysics in Cambridge, England, and at the Massachusetts Institute of Technology, and in 1927 returned to Louvain as professor of astrophysics.

When he returned to Louvain in 1927, Lemaître had already worked out his explanation of the expansion of the universe. This was two years before Hubble went public with his evidence. Lemaître had had some contact with Hubble in America, but the driving force behind his own theory was its ability to satisfy the equations in Einstein's general theory of relativity. When Hubble later published details of the observed red shifts, they provided substantial support for Lemaître's ideas.

Lemaître proposed that the universe had originated at a specific time in the past, from the explosion of a very small, very dense nucleus he termed a 'superatom' or 'cosmic egg', and that it had been expanding ever since. His ideas did not create much of a stir when they were first published; it was not until they were taken up in the 1940s by a physicist with an outstanding gift for popularization – the American George Gamow – that they became known to the world at large.

THE SUN AND ITS ENERGY The first 30 years of the twentieth century were a golden age of observational astronomy; the discoveries made then transformed the science. One consequence was a shifting of astronomy's principal concern from Newton's celestial mechanics, and the workings of the solar system, to astrophysics and the nature and history of the visible universe.

While astronomers were trying to come to terms with these new views of the workings and history of the universe, they were also presented with a discovery that challenged their assumptions about the internal composition of the stars, and the processes that might be taking place within them. The British-born Cecilia Payne-Gaposhkin would later go on to become professor of astronomy at Harvard. But in 1928, when she published the findings with which we are concerned here, she was a Ph.D. student at Radcliffe College, working under the supervision of Henry Norris Russell and occupying a lowly place in the observatory's hierarchy.

At that time, it was still assumed that the Sun was mainly composed of heavy elements, and that the secret of its seemingly inexhaustible supply of energy must lie in atomic reactions involving those. But Payne's studies of the Sun's atmosphere showed it was overwhelmingly composed of hydrogen. This result was so unexpected that it was initially received with scepticism. But soon afterwards, her findings received strong support from two other researchers – the Irishman William McRae and the German Albrecht Unsold – who confirmed that spectroscopic study indicated that stellar atmospheres in general were composed almost entirely of hydrogen. It was a discovery that could not explain the source of the Sun's energy at the time it was made, but it pointed the way to where an explanation might be found.

HYDROGEN INTO HELIUM   In the search for the secret of the Sun's energy, the contribution of two physicists was outstanding. They were the German-born Hans Bethe, working at Cornell University, and the German (Baron) Carl von Weizsacker, working in Berlin. In 1938, they independently identified two processes that were capable of generating huge amounts of energy in the conditions of extreme heat and gravitational pressure existing in the interior of stars. One of these, the carbon–nitrogen cycle, is more relevant to stars rather larger than the Sun. The other, the proton–proton chain, was, they suggested, the source of most of the Sun's energy.

The proton–proton chain is a set of reactions in which hydrogen is turned into helium, with the release of huge amounts of energy as a by-product. In the conditions inside the Sun, it is a self-sustaining process, which has continued for billions of years. Current calculations suggest that more than 90 per cent of the Sun's heat is the product of such reactions, with the carbon–nitrogen cycle accounting for the balance.

Bethe's and von Weizsacker's analysis correctly identified the process responsible for the Sun's enormous energy output; but it was only as a result of further developments in quantum theory, and wartime research into atomic weapons, that the precise nature of the process came to be understood.

The proton–proton chain is an example of *atomic fusion,* in which atoms of one element are rearranged to form the atoms of a heavier element. This is the process that powers the hydrogen bomb. There is a crude sense in which the internal processes of the Sun can be likened to the continuous explosion of millions of hydrogen bombs. It is no wonder that they can cause sunburn, and skin cancer, at a distance of 150 million kilometres/93 million miles.

MENDEL REDISCOVERED The researches of Gregor Mendel in his monastery garden in the 1850s and 1860s would ultimately form the basis of a new science: genetics. But for 40 years after his paper was published, his work was forgotten. When it was rediscovered, early in the twentieth century, its initial effect was to lead biology into a cul-de-sac.

After the publication of *The Origin of Species*, few biologists any longer doubted that present-day species had originated by a process of evolution. Darwin had won that battle decisively. But the gradual process he called 'natural selection' had acquired far fewer supporters, for the very good reason that he had not been able to suggest any mechanism at the cellular level by which it could be brought about. Instead, a rival theory came to the fore: one that explained the development of new species as being the result of *mutations* – changes occurring in cells within the reproductive organs – that gave rise to significant new features in later generations. When Mendel's research was rediscovered, it provided evidence that heredity involved the passing on of little packets of hereditary 'factors'. This encouraged the belief that evolutionary change was a matter of changes in these factors, rather than the selection of already existing variations as a result of external pressures.

GENES AND CHROMOSOMES   Even before the rediscovery of Mendel's work, there had been important discoveries relating to the material involved in sexual reproduction. In the late 1870s, the German zoologist Oskar Hertwig, working with sea urchins, had established that the process of fertilization involved the union of the nucleuses of a sperm cell and an egg cell, and that subsequent development of the embryo consisted of the repeated subdivision of the original combined nucleus. In 1879, another German scientist, the anatomist Walther Flemming, had discovered long, threadlike structures in the cell nucleus, which later came to be called *chromosomes*. In 1903, just after the rediscovery of Mendel's results, an American biologist, Walter S. Sutton, established that chromosomes were arranged in pairs. He suggested that Mendel's 'hereditary factors' were located on the chromosomes, and were passed on in a process of random selection at the moment of conception. In 1909, a Danish botanist, Wilhelm Ludwig Johannsen, coined the term *gene* to denote these hereditary factors, and over the next 20 years the science of *genetics* developed at breakneck speed.

THE COMBINATION OF CHROMOSOMES   Chromosomes within living cells are indeed arranged in pairs, but the number of pairs varies between species. Human beings have 23 pairs. In species that reproduce sexually, the pairs of chromosomes in the cells that are going to become eggs and sperm separate at the moment when these specialized cells are formed, so that an egg and a sperm each contain only one chromosome out of each pair. This process, which is called *meiosis,* involves a complicated shuffling which has been called 'the dance of the chromosomes'. Its significance lies in the random nature of the assortment of genes that finishes up in the nuclei of the egg and the sperm, ensuring that no two individuals – other than identical siblings – will ever receive the same collection of genes from either of their parents. In the case of humans, the fertilized egg ends up with 23 complete pairs of chromosomes – but half of each of the 23 pairs will have been derived from the male parent, and half from the female. The genetic make-up of the offspring will depend on the genes located on the chromosomes the offspring receives from each parent. It is this random element in inheritance that makes sexual reproduction such a powerful mechanism for redistributing large quantities of genetic material through successive generations, producing the enormous variety among individual members of the same species that provides the raw material for the process of natural selection.

THE MECHANICS OF MULTIPLE BIRTHS   Multiple births are of two kinds: those that involve the simultaneous development of two fertilized eggs; and those that result from the splitting of a single egg. Those that arise from a single egg are called *identical.* Those that derive from two simultaneous conceptions are called *fraternal.* Identical twins, triplets, etc., have exactly the same genetic make-up. But fraternal twins, etc., are no more alike in their genetic inheritance than any other siblings that are separately conceived.

The incidence of identical twins is a matter of chance, and averages about 1 in every 250 births, irrespective of the racial origin of the parents. The rate of fraternal (non-identical) twin births, on the other hand, is very much influenced by ethnic factors. For example, in the United States, the incidence of twin birth to Afro-American women is about 1 in 70 and in Caucasian women 1 in 88; but in the case of Chinese women, the figure is around 1 in 300. The odds against multiple births beyond twins mount up in a simple ratio. Thus (in the USA again) while the average incidence of twins is just under 1 in every 90 births, the figure for triplets is 1 in 7,500, and for quadruplets 1 in 650,000.

The odds quoted are all averages. The odds against fraternal births in a specific case depend very much on individual circumstances. They are much more common in the case of women with a family history of multiple births, or with a personal history of multiple births, and in the case of older mothers. Women who use fertility drugs also have a much higher probability of experiencing multiple births.

A single pregnancy may result in a multiple birth involving either identical or fraternal siblings, or a combination of the two, depending on the number of eggs that have been fertilized. For example, triplets may arise in any of the following ways:

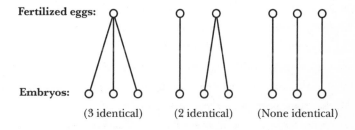

Figure 21. The Mechanics of Multiple Birth
The number of possible permutations leading to the birth of triplets.

GENETICS AND NATURAL SELECTION By the late 1920s, Mendelian genetics seemed to be carrying all before it. Equipped with the concept of mutation, some biologists had come to believe that they had no need of what they considered to be the rather fuddy-duddy ideas of Darwinian natural selection. But a number of Darwin's disciples, led by Ernst Mayr of Harvard, mounted a counterattack; by 1940 a compromise position had been arrived at, which became known as *neo-Darwinism*. This was a marriage of Mendelian genetics and Darwinian theory. Since then, the evolution of new species has been accepted as being substantially the consequence of natural selection – especially in small isolated populations – as a result of environmental pressures, operating through Mendelian inheritance. Random mutation has been relegated to a subsidiary role.

NUCLEIC ACID With agreement having been reached on the big picture, attention next turned to the *biochemistry* of evolution: the mechanisms at the molecular level that code the information that has to be carried within the genes in order for hereditary factors to be passed on to later generations. The prize on offer was a big one. What people were looking for was nothing less than the secret of life.

In 1869, a Swiss biochemist, Friedrich Miescher, working at the University of Tübingen, had discovered a substance that seemed to occur in all cell nuclei. He had given it the name of *nuclein*, but it had later become known as *nucleic acid*. Further research established that there were at least two types of nucleic acid. One of these, which had been the subject of a considerable amount of analysis was given the name of deoxyribonucleic acid, or DNA. Two questions now arose: where precisely in the nucleus was DNA located, and what was its function?

A German chemist, Robert Feulgen, answered the first, by establishing that DNA was to be found in chromosomes, and specifically within the genes. The second question was answered in 1944 by a Canadian-American physician named Oswald Avery, and his colleagues at the Rockefeller Institute in New York.

THE IMPORTANCE OF DNA   Oswald T. Avery was born in Halifax, Nova Scotia, in 1877. He was the son of a clergyman who had migrated to the USA when his son was ten years old. He graduated in medicine from Columbia University in 1904, and joined the staff of the Rockefeller Institute nine years later, in 1913.

Avery's team was carrying out research into two strains – rough-coated (R) and smooth-coated (S) – of pneumococci, the bacteria that cause pneumonia. They discovered that if they mixed live bacteria of the R strain with an extract of the S strain, and injected the mixture into a mouse, the mouse's tissue would in due course contain *live* specimens of the S strain. For the non-specialist, the significance was hard to see; but for biologists the result was sensational. The S strain, although it was no longer alive, clearly still contained something that was able to convert the live R strain into S strain – in other words to *change its genetic make-up*. When Avery and his colleagues were able to show that the material involved was pure DNA, it became clear that DNA was central to the reproductive process. Every researcher in the field was put on notice. Whoever succeeded in establishing its molecular structure might possibly be able to answer the question that Darwin had been unable to answer: what was the mechanism of inheritance that enabled hereditary factors to be passed down the generations and made evolution by natural selection possible?

SPLITTING THE ATOM   While biologists were prising open the secrets of the living cell, the physicists were searching for the secrets of the nucleus of the atom. Thirty years after the publication of his special theory of relativity, Einstein's formula expressing the equivalence of mass and energy remained a speculation without experimental support. He had suggested that it might be possible to test it by experiments involving the bombardment of the nuclei of heavy elements. But it was not until January 1939 that the experiment was performed that demonstrated the formula's validity. It was conducted in the laboratory of the Kaiser Wilhelm Institute for Chemistry in Berlin. The scientists involved were an Austrian physicist, Lise Meitner, and two German chemists, Otto Hahn and Fritz Strassman.

Meitner was born in Vienna in 1878, the Protestant daughter of a Jewish lawyer father. Inspired by the example of Marie Curie, she studied physics in Vienna. In 1907, she travelled to Berlin to attend lectures by the originator of quantum theory, the German physicist Max Plank. It was there that she met Otto Hahn. She remained in Berlin, and in due course she became head of the physics department at the Kaiser Wilhelm Institute, and Hahn became head of the chemistry department. By 1939 they had been working together for 30 years. Meitner was 60 and Hahn was a year younger. Strassman, who had recently joined them, was 36.

In January 1939, Meitner, Hahn and Strassman were engaged in a programme of research based on some earlier work by Enrico Fermi, professor of physics at the university of Rome, who had used a newly discovered particle – the neutron – to bombard the nucleus of uranium. One of their experiments yielded a mysterious result. When they bombarded uranium with neutrons, they found they had produced isotopes of barium and krypton. And for every neutron that had been absorbed in the process, *two* new neutrons had been released. This was startling. But even more amazing was the huge amount of energy released as a by-product of the reaction.

Almost immediately after the experiment, Meitner took a train to Holland, supposedly for a week's vacation. The real motivation for her journey was twofold. She believed that she had been marked down for a concentration camp, which was itself more than sufficient reason for leaving Germany. But, in addition, her mathematical training had given her an insight into the implications of the experiment that had for the moment eluded her chemist collaborators. What they had witnessed, she believed, was the splitting of the uranium nucleus into two more or less equal halves. The huge amount of energy released seemed to her to represent a conversion of mass into energy of the kind postulated by Einstein's formula; and she felt an urgent need to warn people outside Germany of the threat that this represented. While she was in Holland, negotiating a Swedish visa, her two colleagues published a report on the experiment, which as a result became known as the 'Hahn–Strassman Experiment'. But they were unable to offer any explanation of the result.

THE ATOMIC BOMB Hahn and Strassman's paper created a fever of excitement in the scientific community. By the time Meitner's plane touched down, the experiment had already been repeated in Paris by Marie Curie's daughter Irene and her husband Pierre Joliot. In Stockholm, Meitner shared her thoughts on nuclear fission with her nephew, the physicist Otto Frisch, who passed them on to his father-in-law, the Danish physicist Niels Bohr. Bohr was on a visit to the United States, which gave him the chance to discuss the implications with physicists there, including Einstein himself. Within ten days, the experiment had been repeated at Columbia University, at the Johns Hopkins Laboratory, and at the Carnegie Institute in Washington.

Among the team at Columbia was Fermi himself, now a refugee from Italian fascism. He discussed with Bohr the possibility of exploiting the process already observed to initiate a *chain reaction*. This was a reaction in which the radioactive particles produced in the initial moment of fission would, like a firecracker in a munitions factory, disrupt other nuclei in an accelerating reaction, culminating in massive explosion. If such a process were a practical possibility, calculations suggested that a pound of uranium might yield an explosion equivalent to 40 million pounds of TNT.

The question now arose, if such a reaction were possible, why had it not occurred in the laboratories where atomic fission had been demonstrated? The answer was that the uranium used had been composed of three different kinds, only one of which – uranium 235 – was capable of fission; this represented less than 1 per cent of the mass of the samples used.

In March 1940, six months after the Germans invaded Poland, the Columbia team was supplied with their first samples of pure uranium 235, and they were able to confirm that it was this isotope that was the source of the fission product. Einstein himself had said, only a year or so earlier, that he was 'sure, nearly sure, that it will not be possible to convert matter into energy for practical purposes for a long time.' But by October 1939, he had felt sufficiently unsure to write to President Roosevelt. He told him that it might be possible to achieve a nuclear chain reaction 'in the immediate future'; adding that it was 'conceivable ... that extremely powerful bombs of a new type might thus be constructed'. The President's response was to create the body that became the National Defense Research Council.

Knowing that the Germans were also researching atomic weapons, the British government initiated its own programme. By the summer of 1941 the British team were certain that such weapons were a practical possibility. In November, the National Defense Research Council reported to the President its conclusion that, 'If all possible effort is spent on the program, we might expect fission bombs to be available in significant quantities within three or four years'. On 6 December, the government decided upon an 'all-out' effort to develop an atomic bomb. The next day, the Japanese attacked Pearl Harbor. Physics was suddenly in the front line.

THE FIRST NUCLEAR REACTOR The 'all-out' effort that the United States embarked upon was the most ambitious exercise in applied science the world had ever known. At its peak, the workforce numbered 125,000 people. Many of these were employed in the manufacture of the raw materials of the bomb in factories at Oak Ridge, Tennessee, which at one stage were consuming $\frac{1}{7}$th of all the electricity generated in the country.

While these facilities were being created, some of the world's most gifted nuclear physicists addressed themselves to the task of proving the feasibility of a controlled chain reaction. 'Controlled' in this context meant a 'slow', self-sustaining reaction, in which the number of neutrons being produced exactly equalled the number escaping.

One problem the group faced was that the speed of the neutrons, and the expanse of subatomic space they moved through, was so great, that most of them would escape before they could collide with one of the atoms they were meant to be splitting. It was therefore necessary to find a 'moderator': something that would slow the neutrons down, and increase their chances of being involved in a collision. The moderator chosen was pure graphite; and a 'pile' of graphite blocks was constructed in a squash court under the Stagg Field stadium of the University of Chicago.

Another problem the group faced was that, if they proved unable to control the process they had started, an unstoppable chain reaction might eliminate, not only them, but a substantial piece of the university too. In theory there was a margin of time available for inserting the control rods used to slow the reaction if it threatened to get out of hand. But the margin was small; and it was with some care that they checked their calculations again before throwing the switch to start the nuclear reaction. The Geiger counters, and the group's collective pulse, began to race.

At 3.20 p.m. on 2 December 1942, Fermi was able to announce to his colleagues that they had achieved the world's first self-sustaining chain reaction. It fell to a Nobel laureate, Arthur H. Thompson, to make the call informing the President of their success: 'The Italian navigator has just arrived in the New World,' he told him. The Italian navigator was not alone. Everyone was now in a New World.

NUCLEAR WARFARE The world's first *uncontrolled* chain reaction followed. It took place at 5.30 a.m. on 16 July 1945, in a thunderstorm at Alamogordo Air Base in New Mexico, when a bomb containing about 9 kilograms/20 pounds of uranium 235 and plutonium was detonated on top of a steel tower, in a flash that reached 50 million degrees Celsius/100 million degrees Fahrenheit, and with an explosive force equivalent to 20,000 tons of TNT, leaving a crater 0.8 kilometre/0.5 mile wide. Three weeks later, at 8.15 a.m. on 6 August, its twin was detonated above the Japanese city of Hiroshima, creating a firestorm that incinerated 80,000 citizens, and left a wasteland where a great city had stood. On 9 August, to underline the message, a similar weapon was detonated over the city of Nagasaki. It ended the Second World War.

THE NUCLEAR ARMS RACE The scientists who campaigned against nuclear weapons were wasting their breath. Having mastered the technology of atomic *fission,* those countries that could afford to do so were determined to acquire the even more potent technology of atomic *fusion.* During the 1950s, first America, then Russia, then the United Kingdom showed off their scientific muscle by detonating hydrogen bombs.

Hydrogen bombs achieve their effect by converting hydrogen into helium – the process that provides the heat output of the Sun – and they use Hiroshima-type bombs as triggers. The Russians managed the biggest bang. On 20 October 1961, on the far northern island of Novaya Zemlya, they detonated a device that converted 3 kilograms/7 pounds of mass into free energy. The explosive force created, expressed in the shorthand employed to measure these events, was 57 megatons: 3,000 times greater than the blast that wiped out Hiroshima. It's called 'applied science'.

LINUS C. PAULING   The greatest chemist of the twentieth century was born in Portland, Oregon, in February 1901. His name was Linus C. Pauling, and he was the son of a druggist, and the eldest of three children. His father died when his son was 9 years old. Pauling's interest in chemistry dated from the day he was given a chemistry set at the age of 13. He attended Oregon Agricultural College; and in his third year his teachers were so impressed with his ability that he was paid a salary to teach the second-year course in quantitative analysis he had just completed himself. It was a welcome offer, as he was by then supporting his mother, who was seriously ill.

He graduated in 1922, and proceeded to the California Institute of Technology (CalTech). His doctoral thesis there was concerned with the structure of crystals, as revealed by *X-ray crystallography*, a powerful new procedure invented in 1912 by the German physicist Max von Laue. X-ray crystallography is the analysis of crystal structure using a technique known as X-ray diffraction.

Pauling gained his Ph.D in 1925; joined the faculty at CalTech; and was appointed a professor in 1927. Four years later, he revolutionized atomic theory with his analysis of the nature of the chemical bonds by means of which atoms join together to form molecules. This analysis, which was contained in an article in the *Journal of the American Chemical Society*, introduced into chemistry ideas from the new science of quantum mechanics, treating electrons as wave forms rather than particles. It explained chemical bonding as the joining in pairs of electrons from individual atoms. In 1939 he summarized his ideas on molecular structure in a book entitled, *The Nature of the Chemical Bond,* which became a defining text of twentieth-century chemistry. In 1954, he was awarded the Nobel Prize for chemistry. In 1962, he became the second person after Marie Curie to win *two* Nobel Prizes when he received the Nobel Prize for peace for his efforts to achieve nuclear disarmament.

One of Pauling's many contributions to the understanding of molecular structure was his explanation of the characteristics of materials in terms of the chemical bonds between individual molecules. So, for example, he was able to show that the hardness of diamond compared with graphite (both forms of pure carbon) was the consequence of the different way in which the carbon atoms bonded together.

In the 1940s, he and his colleagues turned their attention to the large carbon-based molecules – proteins and amino acids – that characterize living matter. Proteins provide the structure, and drive the body processes, of plants and animals. Amino acids are the raw material out of which proteins are made. Examples of structural protein include fingernails and muscle. An example of a process-directing protein is the enzyme amylase in saliva, which turns starch into sugar. Some of these proteins have enormously complex molecules, weighing thousands of times as much as a molecule of water; and every species of living creature manufactures unique proteins that other species cannot directly utilize. This is why animals have to digest and break down the protein they get from eating plants, and/or other animals, in order to be able to reuse the amino acids that the proteins contain.

In May 1951, the Caltech team created a sensation when they published seven papers describing the structure of the proteins in silk, feathers, hair, and a number of other organic materials. But in the race for the most glamorous prize of all – the decoding of the structure of DNA – they were pipped at the post by a group of individuals based in the English universities of London and Cambridge.

THE STRUCTURE OF DNA By the beginning of the 1950s, it was clear that:

1. The thread-like chromosomes within the sperm and the egg carried the hereditary factors that determined the character of the developing embryo.
2. The hereditary factors themselves were the genes that were strung out along the chromosomes,
and
3. The 'instructions' carried in the genes were contained in chemical form in the molecular structure of deoxyribonucleic acid – DNA.

It was also clear that:

A. Someone, very soon, was going to establish the structure of DNA.
B. The understanding thereby acquired would transform the study of genetics,

and

C. Whoever made the discovery would be assured of
undying fame.

The inevitable breakthrough came in 1953; and the people who
hit the jackpot were an English physicist turned molecular
biologist named Francis Crick, and an American biochemist
named James D. Watson.

X-RAY CRYSTALLOGRAPHY Francis Crick was born in
Northampton, England in 1916. He studied physics at London
University, and worked on radar during the Second World War.
In 1946, he attended a lecture given by Linus Pauling, which
opened his eyes to the possibilities of original discovery in
molecular biology. This led him to apply to do research in biology at Cambridge; in 1949, when he was 33 years of age, he
was taken on by the Medical Research Council Unit at the
University's Cavendish Laboratory.

James Dewey (Jim) Watson was born in Chicago in 1928,
and was a child prodigy. He enrolled in the University of
Chicago at the age of 15, graduated when he was 19, and
3 years later obtained a Ph.D. from the University of Indiana.
While preparing for his Ph.D., he read a book entitled *What Is
Life?* by the Austrian physicist Erwin Schrödinger, which persuaded him that the study of the gene offered prospects of
exciting discovery. In 1951, he attended a conference in Naples,
where he met Maurice Wilkins, a 33-year-old New Zealand-
born British physicist, who had worked on the atomic bomb in
America, but who had turned away from nuclear physics in
disgust at the consequences of that work.

Like Watson, Wilkins had been inspired by Schrödinger's *What is Life?* He was now working in the Medical Research Council's Unit at London University's King's College on the structure of large organic molecules, using the same technique of X-ray diffraction analysis as Pauling's team at CalTech. Wilkins' description of his work reinforced Watson's interest in the subject, and the latter applied, and was accepted, to do research at the Cavendish Laboratory. He arrived in Cambridge shortly after his 23rd birthday, establishing an immediate rapport with the 35-year-old Crick.

Crick and Watson were determined to investigate the structure of DNA, but they were discouraged by their superiors, who were aware of research already in progress at King's College. The work at King's was supposedly a team effort by Wilkins and a 30-year-old British chemist named Rosalind Franklin; but it was handicapped by a personality clash between them. Franklin was a highly skilled crystallographer. Crystallography is a demanding technology, based on the technique of X-ray diffraction, which is essential in the investigation of the structure of large molecules. It was a technology in which neither Crick nor Watson possessed any expertise. In its absence, they did the best they could with the only alternative available to them – model-building. But without the clues that crystallography could provide, they were unable to make real progress. And it could only be a matter of time before Pauling's team at CalTech, who were skilled in both model building and crystallography, would hit upon the correct answer. It was Pauling's book, *The Nature of the Chemical Bond,* that was Watson's bible in his attempts to construct a plausible model. Frustratingly, the director of the Cavendish Laboratory,

Lawrence Bragg, and the head of its Crystallography Unit, Max Perutz, were expert crystallographers. But they were insistent that the goodwill of the Medical Research Council must not be put at risk by duplicating research it was funding at King's.

It was not a question of determining the chemical composition of the DNA molecule. It was well known by this time that it was built up out of a succession of *bases*, which came in four kinds: thymine (T), guanine (G), cytosine (C), and adenine (A), attached in pairs to a phosphate-sugar backbone. But no one knew what form the backbone took, or how the base pairs were attached to it. Without answers to these questions, there could be no real understanding of the detailed mechanism of inheritance; and certainly no possibility of applying theoretical knowledge to real-life problems such as inherited disease.

Ironically, the breakthrough came soon after Crick and Watson attended a seminar in which they seem to have misunderstood a presentation that Franklin made on her research. They hurried back to Cambridge, constructed a model, and invited the London pair to view it – only to have Franklin shoot their ideas down in flames. Not long after, with Wilkins' help, Watson got sight of an X-ray crystallograph of astonishing clarity that Franklin had produced. As soon as he saw it, he felt sure he knew how it should be interpreted.

Back in Cambridge, he and Crick now obtained permission to employ the services of the laboratory's machine shop to construct a large-scale model of the molecule. After a frenetic and brilliant five weeks of trial-and-error, the model was ready to be unveiled. It took the form of a *double helix,* a long winding

THE DOUBLE HELIX

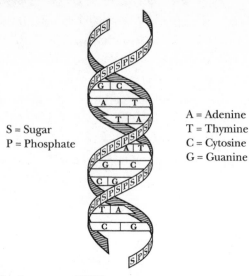

S = Sugar
P = Phosphate

A = Adenine
T = Thymine
C = Cytosine
G = Guanine

Figure 22. The Structure of DNA

The main features of the DNA molecule are:

a.   A 'spiral staircase', with sugar-phosphate 'rails', forming a 'double helix'.

b.   A long succession of 'treads', made up of interlocking base pairs of either adenine (A) and thymine (T) or guamine (G) and cytosine (C).

N.B. 1.   A 'gene' is a length of DNA, usually several thousand bases long.

2.   The sequence of bases in the genes encodes the information needed to manufacture the proteins that determine the anatomy and physiology of the living creature.

3.   Genes are located on chromosomes, which are long strips of mixed DNA and protein.

4.   In sexual reproduction, the gene sequence in the fertilized egg is different from either parent, making every individual offspring unique.

5.   In identical siblings, the gene sequence differs from both parents, but is identical in each sibling, because they are the result of the splitting of one fertilized egg.

6.   When cells split and multiply during the growth of the embryo, the base pairs open up like a zip fastener. Every half of a base pair has the ability to 'grow' a new partner in its new home, ensuring that the 'information' in every cell is the same.

ladder in which the rungs were innumerable sequences of base pairs: TA, CG, AT, TA, GC, and so on.

On 7 March 1953, it was shown to their colleagues. On 25 April, a short, modestly worded paper in the journal *Nature,* entitled *Molecular Structure of Nucleic Acids,* informed the world of one of the most significant breakthroughs in the history of science. It was somewhat overshadowed by the first ascent of Everest six weeks later. But its significance soon sank in, and it initiated an explosion of genetic research and discovery that still continues.

In 1962, Crick, Watson, and Wilkins shared the Nobel Prize for medicine. Rosalind Franklin's name was not mentioned. She had died of cancer in 1958, aged 37, a victim – like Marie Curie – of the radiation she had spent her days with. Nobel Prizes are not given posthumously.

'BIG BANG' OR 'STEADY STATE'? The urgency of wartime needs that led to the development of the atom bomb had yielded a new understanding of the processes at the heart of the atom. As the world returned to peacetime normality, physicists turned their attention to problems at the opposite end of the scale: the structure and origins of the universe.

The hypothesis put forward in the 1920s by Lemaître – that the universe had originated in the explosion of a 'cosmic egg' – was now more widely known, and had many supporters. But unlike the accepted fact of an expanding universe, it still lacked supporting evidence. In the late 1940s, it acquired a new champion in the person of an American physicist named George Gamow.

Gamow, the grandson of a tsarist general, was born in Odessa, in the Ukraine, in 1904. He obtained his Ph.D. from Leningrad University, and later held posts in universities in Western Europe. He moved to the United States in 1934, and made his home there. He was a teacher at George Washington University until 1956, when he joined the faculty at Colorado University.

While he was a member of the faculty at George Washington University, Gamow acted as a consultant to the Applied Physics Laboratory at Johns Hopkins University, and it was there he made the acquaintance of a Ph.D student named Ralph Alpher. In 1948, Gamow and Alpher published a paper in *Physical Review,* in which they discussed the conditions that would have existed at the heart of a densely concentrated 'superatom' of the kind that Lemaître had suggested might have exploded to create an expanding universe. The paper marked the formal introduction of what we now know as the 'big-bang' model of the origin of the universe.

Opposed to Gamow and Alpher was a group at Cambridge University in England that included an equally gifted astrophysicist and science popularizer named Fred Hoyle. In the year that Gamow and Alpher published their article, Hoyle and his co-theorists – Herman Bondi and Thomas Gold – advanced a rival explanation, which they christened the 'steady-state' hypothesis. They suggested that the observed expansion was the result of the continuous creation of new matter in the space between the galaxies, pushing them further apart. In this model, the universe would have had no beginning; it would have the same general structure wherever one might look; and it would always have looked as it did in the present day. It was Hoyle who coined the phrase 'big bang'. He meant it dismissively. But the name stuck, and its supporters wore it as a badge of pride. Nevertheless, both rival explanations were still the purest of hypotheses.

In another 1948 paper, Alpher and a colleague named Robert Herman had predicted that, if the universe had originated in a 'big bang', there would be faint, very cold, 'background' radiation lingering behind, like an echo. In 1965, two workers at the Bell Laboratories at Holmdel, New Jersey – Arno Penzias and Robert Wilson – accidentally discovered microwave radiation that was of just the temperature and strength one would expect to exist if a big bang had occurred; it seemed to come from every direction in space. It was a discovery for which they received a Nobel Prize, and it marked the end of the steady-state hypothesis. As new telescopes were built, and astronomers were able to peer further into the distance – and back in time – they were able to see a more and more crowded universe, and progressively younger galaxies.

In the 40 years since Penzias and Wilson discovered the background radiation, the big bang has become the accepted explanation of the origin of the universe. The mystery remains as deep as ever; but the debate has moved onto new ground: what came before the big bang, and whether 'before it' can have any meaning. One idea that has been advanced is that the history of the universe may be a story of successive expansions and contractions, rather than one single expansion from a unique big bang.

HOW MANY STARS? One answer that science has had to keep revising is the answer to the question 'How many stars are there in the sky?' If one means stars visible to the naked eye, then the answer is what it has always been: 6,000. But as telescopes have become more powerful, so the number of *known* stars has increased. There are 400 billion stars in our Galaxy, which is only an average sort of galaxy. And there are around 100 million other galaxies we know of. So if you want a ballpark figure, you could try 40 million, million, million. Give or take a star or two.

LIFE EXPECTANCY If you want to live a long time, it's best not to be small. At least that seems to be the rule when comparing species. From the mayfly at one end of the age scale, to the pine tree at the other, there is a marked correlation between size and longevity.

The following are average life expectancies, in the wild, for some species of vertebrates (animals with backbones). Animals in a protected environment may live much longer.

	Years
Mammals:	
Vole	1
Grey squirrel	5
Wolf	10
Asian elephant	40
Reptiles:	
Rattlesnake	10
Nile crocodile	40
Fish:	
Trout	3
Sturgeon	30
Birds:	
Chickadee	2
Golden eagle	20

Life spans can be much longer than these figures. An Asian elephant can live for 80 years. A female sturgeon would consider herself in her prime at 100. It's also a good rule not to rush about: Galapagos giant tortoises are still going strong – if that's the word – when they are 120.

If you keep absolutely still, you can really notch up the years. There is a bristle-cone pine in Nevada that was already 2,000 years old when King Nebuchadnezzar of Babylon was just a glint in his mother's eye.

THE GENES WE SHARE   A term that haunted twentieth-century discussions of human origins, and still crops up occasionally, is 'the missing link'. It would be difficult to think of another short phrase that has caused so much misunderstanding. The missing link was an intermediate stage between apes and humans that was supposed to be the essential piece of evidence needed to prove the truth of humanity's descent – or ascent – from the apes, and without which Darwin's theory of evolution by natural selection was mere empty speculation.

The only thing wrong with this proposition was that neither Darwin nor any other biologist had ever suggested that human beings *had* evolved from apes, any more than anyone had ever suggested that French had evolved from Spanish. What Darwin said, and what genetic evidence confirms, was that apes and humans share a common ancestry, just as French and Spanish do.

The closeness of the family relationship is revealed by comparisons of the genetic make-up of the species involved. Recent studies suggest chimps and humans share 98.4 per cent of their genes. The figure for gorillas and humans is 97.7 per cent. If humans are 98 per cent chimpanzee, it is equally true to say that chimpanzees are 98 per cent human.

The idea of a 'missing link' between apes and humans only arises if one thinks of evolution as a *ladder* from 'lower' to 'higher' forms of life. Darwinian theory doesn't recognize species as being higher or lower. They are just different. Darwin saw evolution as a *branching tree,* with apes and humans as twigs on the same branch.

Comparison of the shared genetic inheritance between present-day species yields the following picture of this process of repeated divergence:

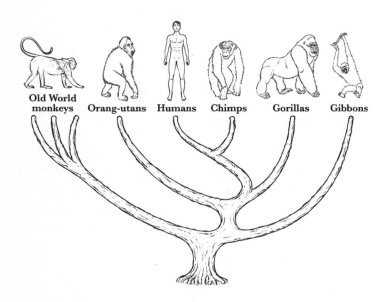

Figure 23. The Primate Family Tree
The successive divergence of monkeys, apes and humans from a shared ancestry.

HUMANOID HISTORY   It is now generally accepted that the split between the line that led to modern humans and the line that led to chimpanzees occurred in Africa about 6 or 7 million years ago, when two groups of an ancestral species got separated, and developed along different lines. But because of the small number of fossil remains, no one can say what the subsequent line of human evolution was, and which of the ten or so species of *homo* that have so far been discovered represent stages in human evolution.

The following table shows the relationship in time of some of the better-known fossil species, but the table does not imply any other kind of relationship. The dates are only approximate. We can be confident that our picture of human evolution will become clearer over the next 30 years or so, as more fossils are discovered, and studies of past and present-day DNA are extended. But it is by no means certain that we will even then be close to establishing humanity's ancestral line.

Species	First appearance (years BP – before the present)
Australopithecus afarensis	4,000,000
Homo habilis	2,500,000
Homo erectus	1,600,000
Homo neanderthalensis	200,000
Homo sapiens (modern humans)	120,000

Australopithecus afarensis – 'southern apeman from Afar (in Ethiopia)' – was an upright-walking, rather ape-like hominid, that lived in East Africa.

Homo habilis ('handy man') – the earliest humanoid – was a toolmaking species that lived in southeast Africa, and probably had the beginnings of speech. There was a similar species named *Homo ergaster*.

The earliest remains of *Homo erectus* ('upright man') are found in Africa, but fossils less than 200,000 years old have been found in Java, China, and the Caucasus. *Erectus* was a fire-user, and a sophisticated toolmaker, with a brain size approaching that of modern humans; and, probably possessed a well-developed power of speech.

Homo neanderthalensis ('Neander Valley man') was a strongly built species that lived in Central Asia, the Middle East, and Europe at the same time as *Homo sapiens* (and had a bigger brain). There is no evidence that the two species interbred.

Neanderthals became extinct about 30,000 years ago. Whether *Homo sapiens* had anything to do with this is not known.

Homo sapiens ('thinking man') is our own species, and includes all living humans. There is remarkably little genetic variation within the species throughout its distribution, and it has remained essentially unchanged for the past 100,000 years.

THE CULTURAL TIME SCALE The timescale on the previous page is a biological one, concerned with the succession of humanoid *species*. Archaeologists apply a timescale of their own to the past 2.5 million years. This is concerned with the succession of *cultures,* and it is defined primarily in terms of artefacts: tools, weapons, utensils, etc.

Oldowan (2.5 to 1.5 million years BP). East Africa. The earliest known stone toolmaking culture.

Acheulian (1.5 million to 200,000 years BP). Africa, Near East, and Europe. A prehuman culture characterized by the use of hand-axes and other roughly flaked tools of flint, chert, quartzite, etc. Associated with *Homo erectus* and *Homo ergaster*, among others.

Mousterian (200,000 to 35,000 years BP). Africa, Near East, and Europe. More developed hand-axes, and flaked tools made from prepared cores. Associated mainly with *Homo neanderthalensis*, but also with early *sapiens*.

Late Paleolithic (Old Stone Age; 40,000 to 12,000 years BP). Africa, Asia, and Europe. Much more sophisticated weapons and tools. Cave and rock art, jewellery and personal ornament.

Neolithic (New Stone Age; 12,000 to 6,000 years BP). Virtually worldwide. Sophisticated stone tools, ceramics, and textiles. Following the ending of the last ice age, the agricultural revolution in the Middle East, and in south and east Asia initiated a new way of life; but elsewhere most cultures remained hunter-gatherers.

GENES AND GENOMES   *Genes* are the packages of information carried on the chromosomes in the cells of every plant and animal. They are the 'software' that directs the vital processes of the living creature. In the embryo, and in the immature plant or animal, they provide the instructions that guide the development of the features that the adult form will exhibit. Every cell contains the same collection of genes, arranged along the chromosomes. In creatures that propagate sexually, every cell other than the sex cells contains two copies of every chromosome; the sex cells contain only one copy.

The *genome* is the name given to the entire sequence of genes (and the base pairs that make up the genes) in a particular species. Some species have quite short genomes, containing a small number of genes. The first genome to be sequenced, in 1977, was that of a virus, which contained only 5,386 letters (base pairs) of genetic code. The human genome, which was sequenced in 2001, contains 22 chromosomes, plus one sex-determining chromosome, making 23 in all. These carry 30,000 genes, which in turn contain 3,000 million letters of genetic code. Even so, the genes account for only 5 per cent of the total amount of DNA in a human cell. The rest – at the moment – is labelled 'junk DNA', because no one knows whether it serves any purpose. It is quite likely that future research will reveal that it is anything but 'junk'.

The 'human genome' that was sequenced in 2001 was not the genome of a real-life human being. It was an anonymous 'average human'. The Human Genome Project that produced it became possible because scientists had access to equipment that could chop DNA into short pieces for distribution to large numbers of researchers, and powerful computers that could handle the number-crunching involved.

A SMALL HUMANOID It can be unsettling to discover relatives one never knew one had. In 2004, a team of Australian palaeontologists discovered fossil remains of what they took to be a previously unrecorded variety of humanoid on the Indonesian island of Flores. 'Humanoid', rather than 'human'

THE HUMAN GENOME

Chromosome	Number of genes (rounded off)	Number of base pairs (millions rounded off)
1	3,000	280
2	2,300	250
3	2,000	220
4	1,300	200
5	1,600	200
6	2,000	180
7	1,400	160
8	1,100	150
9	1,300	140
10	1,400	140
11	2,100	150
12	1,700	140
13	700	120
14	1,100	110
15	1,100	100
16	1,100	100
17	1,600	90
18	800	90
19	1,500	70
20	900	70
21	300	50
22	300	50
	30,600	3,060

Plus chromosome 23: either an X chromosome (1,200 genes) or a Y chromosome (200 genes). In females, chromosome 23 is always an X. In males, it may be an X or a Y. The sex of the offspring depends on whether it receives an X or a Y from the male parent. If it receives an X from both parents, it is female; if it receives an X from its mother and a Y from its father, it is male.

seemed to be the correct term in this case, since, in the opinion of its discoverers, the creature in question did not appear to be in any sense a modern human.

The remains, which included a complete skull, were found 6 metres/20 feet below the floor of a cave called Liang Bua, in Central Flores. They appeared to belong to a 1-metre/3-foot-high creature, of human rather than ape-like appearance, to which their discoverers gave the name of *Homo floresiensis*. For the benefit of the media, they nicknamed their new species 'the hobbit'. Carbon 14 dating yielded an age of about 18,000 years. Fragmentary remains of six other individuals from other strata in the cave yielded dates ranging from about 70,000 years ago to possibly as recent as 12,000 years ago.

The capacity of the complete skull indicated a brain size of about 400 ml: larger than that of the average chimpanzee, but much smaller than the average of around 1,300 ml for modern humans. But given that there was only one complete skull, that brain size in mammals is related to body size, and that there is little correlation between brain size and intelligence in modern humans, this particular specimen provided no clue to the abilities of these new recruits to the human family. Stone tools, animal bones, and signs of fire were found nearby, and appeared to be associated with the remains.

The finding of a pigmy human was not particularly surprising. There are small-stature modern humans in Africa, and it is a recognized feature of evolution in isolated populations that both giant and dwarf species can arise in a very short period of time, geologically speaking. Flores was biologically isolated for at least 1 million years in prehistoric times, and giant rats and pigmy elephants were already known to have evolved there. But to see the principle displayed in the evolution of what appeared to be a close relative of *Homo sapiens* still came as a shock. And what caused further surprise was the late date attributed to the remains.

The most recent date previously associated with any humanoid species was that attributed to Neanderthal man, *Homo neanderthalensis*, a species that appears to have become extinct 30,000 years ago. The discovery of another close relative that was apparently grilling steak around 12,000 years ago raised the question of just when *that* species – if it *was* a new species – died out. Its discoverers could not resist hinting at the possibility that it might not have done so. As it happens, Flores was devastated by a huge volcanic eruption about 12,000 years ago, and that may well represent the date when the small people of Flores consumed their last entrecôte. But every corner of the world has its tales of 'little people', and even anthropologists are allowed to dream. Some experts, however, remain unconvinced that the discovery represented anything other than a pigmy modern human of a kind already known on the island.

The discovery did not, in any case, call for any redrawing of humanity's family tree. But it was a reminder of how fragmentary our knowledge of human evolution is, and how many surprises still await us.

THE RATE OF EVOLUTION   When Darwin formulated his theory of evolution by natural selection, he was still under the spell of his friend and mentor, the Scottish geologist Charles Lyell. Lyell was the advocate of a concept called *Uniformitarianism*, which was a clumsy term for the idea that the history of the Earth had been a matter of slow gradual change.

Darwin fell for this idea, and it was woven into every page of *The Origin of Species*. Uniformitarianism was an understandable response to the opposing school of *Catastrophism*, which explained the history of the Earth in terms of cataclysmic events such as the biblical Flood. But it was an ideology, not a theory, and it embodied only a partial truth.

We know much more about the history of life on Earth than Darwin knew; without detracting from his achievement, we can see that his insistence on gradual change as the exclusive motor of evolution was an aberration. For most of the time, evolution does proceed by small steps. But every so often the process is speeded up by a shock event such as the asteroid impact at the end of the Cretaceous Period. But in the absence of violent changes in the environment, living forms may remain essentially unchanged for millions of years. The dust raised by the battle between the competing ideologies of Uniformitarianism and Catastrophism has settled, and the truth – as might have been expected – can be seen to lie in the middle ground between them.

MASS EXTINCTIONS The extinction event at the end of the Cretaceous Period that witnessed the demise of the dinosaurs was just one of many such events since life first appeared on the Earth. Each time, entire groups of plants and animals, some containing thousands of species, have vanished, never to reappear. And each time, the surviving groups have exploded in a riot of diversification, to fill the gaps left open. Some of these extinctions have undoubtedly been the result of cataclysmic events such as asteroid impacts, or huge volcanic eruptions. But some seem to have been more drawn out: the result of marked but gradual climate change, rather than disturbances in the Earth's fabric. Evidence continues to accumulate of long-term cycles in the Earth's climate that are related to the planet's motion in space, sometimes initiating great ice ages, at other times bringing tropical heat to most of the Earth. As our understanding of such matters grows, it seems likely that we will have to credit climate change with a leading role in extinction of many formerly widespread species that have left their traces in the fossil record.

THE PASSENGER PIGEON   Not all extinctions are the consequence of natural disasters. Two hundred years ago, passenger pigeons were probably the most common birds in eastern and central North America. They congregated in huge flocks; and their migrations darkened the sky for hours at a time. In the markets of New York City, their carcasses sold for a penny apiece. The ornithologist Alexander Wilson estimated that one flock he observed contained over *2,000 million* birds. Little more than 100 years later, on 1 September 1914, the last passenger pigeon on Earth – a bird named Martha – died in Cincinnati Zoo.

The species had fallen victim to the deadliest predator on the planet: modern humans. No one set out to exterminate the passenger pigeon; no one at the beginning of the nineteenth century would have thought it conceivable that so abundant a species *could* have been eliminated in such a short space of time. But two highly effective technologies – the electric telegraph and the rifle – enabled hunters to ambush the flocks as they migrated; and mere numbers proved no protection.

ENVIRONMENTAL DESTRUCTION   The story of the passenger pigeon carries a terrible message. It was not the first example of the adverse effect that technological advance can have on the environment, and the life forms that share it with us. Ever since people began to clear forests and plant crops, they have exposed vulnerable soils and created deserts. But while their numbers were small, and their technology limited to hand tools, the damage they did was not of great significance in the overall balance of nature. During the nineteenth century, as machinery proliferated and numbers increased, the damage mounted, and pollution became widespread.

Round about the middle of the twentieth century, humanity's assault on the planet's health entered a new phase. The pollution and destruction of the atmosphere by the waste products of air travel; the pollution of the seas by industrial effluent; the increase in over-fishing and over-cropping; the release into the environment of thousands of synthetic chemicals no species had ever met – all these raised the possibility of a mass extinction of a totally new kind: the destruction of a substantial portion of the world's life forms by the action of just one species.

In recent years, the sources of damage have multiplied; and the pace of environmental and species destruction has increased. There is no longer any question as to whether a major extinction is imminent. It is already in progress. It is now just a question of how great the destruction is going to be.

THE BLIGHT OF BAD SCIENCE   And while we are telling sad stories, let us consider the case of Trofim Lysenko, geneticist. He was born in Karlovka, in the Ukraine, in 1898. He was a graduate of the Kiev School Institute of Agriculture, and he was an example of the kind of scientist whose vanity and lust for power obliterate any lingering regard they might have for evidence or experimental proof. Whether he believed the rubbish he advocated is unclear. What *is* clear is that his rejection of Darwinian natural selection and Mendelian genetics set back the study of biology in Russia by a generation.

Lysenko came to Stalin's notice during the Second World War, when the need to increase agricultural production was desperate. His basic proposition was that inheritance did not reside solely in the gene, and it was possible to change the genetic character of plants such as food cereals by subjecting them to a changed environment. This implied that one could significantly increase crop yields between one generation and the next. He even suggested it might be possible to grow wheat from rye grains. This was music to Stalin's ears, and in 1940 he made him director of the Institute of Genetics of the Academy of Sciences, a post he held for the next 25 years. For most of this time his ideas dominated Soviet biology.

Geneticists who concealed their doubts were able to continue working. Of the dissenters, the more fortunate merely lost their jobs. The less fortunate ended up in prison, or dead. Lysenko's authority was diminished by Stalin's death in 1956, but he held on to his position for another nine years. It was not until 1965 that Russia finally rid itself of this malign influence, who had done so much damage to its science, and to its scientific standing in the eyes of the world.

This story of Stalin and his tame biologist carries a moral for all those who would bend science to meet the needs of ideology. The health and wealth that developed nations enjoy today is the product of advanced technology; and that technology is the product of five centuries of scientific discovery. There is no law that says that advance is inevitable: that once it has got under way, it is bound to continue. We have seen over and over again in the course of this book how closely scientific progress is linked to the way societies are organized and the way societies think. Stifle the spirit of enquiry, and you can bring scientific progress to a halt as more enlightened nations move ahead. People who, in our own day, are tempted to censor the teaching of evolutionary biology, need to be aware that they are playing with fire. The country that votes to stifle science is voting for economic stagnation and national decline.

DISTANT ALIENS Since we are talking about bad science, perhaps this is an appropriate place to say a few words about aliens. A few words are all that are needed. The nearest star to the Earth is called Proxima Centauri. It's 4.3 light years away. That's 40 million million kilometres/25 million million miles. Think about it....

PARADIGM SHIFTS Many people think of scientific discovery as a process of gradual accumulation of new knowledge, which is added to a pile of existing knowledge. This is what one might call the 'sand castle' view of science, which sees individual scientists, no matter how eminent and adventurous, as children digging on a beach, adding their contribution to the pile of sand that has already accumulated. This might describe 98 per cent of what we call scientific advance. But we need another image to convey the nature of the other 1 or 2 per cent.

A recurring theme of this book has been the shock of new ideas, and the readjustment of scientific thought they bring about. This process of readjustment was the subject of a book published in 1962 entitled *The Structure of Scientific Revolution,* by Thomas S. Kuhn (1922–1996), professor of linguistics and philosophy at the Massachusetts Institute of Technology.

Kuhn's thesis was that scientific discovery is for most of the time a process of gradual accumulation of knowledge and understanding within the limits of what he called 'normal science'. But once in a while, a 'new paradigm' – a revolutionary new model – is put forward, which offers a dramatically changed view of the underlying reality a particular science is trying to explain. If the new model proves successful in explaining hitherto mysterious phenomena, a period of upheaval ensues, as scientists try to come to terms with its implications. This leads to a reorientation of the science in question, which Kuhn called a 'paradigm shift'. In due course the new paradigm meets with general acceptance, and there follows a period of exceptionally fruitful enquiry, which may last for two or three centuries, as scientists explore the territory the new view has opened up.

Paradigm shifts need not be destructive. To return to the children on the beach, a major scientific breakthrough need not mean the flattening of the sand castle. It would be more like someone saying, 'Why don't we build an ocean liner instead?' If it seems like a good idea, it generates a burst of enthusiasm, and construction, that the original plain sand castle could never have produced.

The following is just a selection of some notable scientific revolutions of the past 600 years. All of them represent paradigm shifts of the kind Kuhn had in mind. And all of them were followed by a quickening of the pace of scientific discovery that continued for a long time.

The Sun-centred Model of the Solar System
The Law of Universal Gravitation
The Periodic Table of the Elements
Evolution by Natural Selection
The Planetary Model of the Atom
Special and General Relativity
The Expanding Universe
The Structure of DNA
Plate Tectonics

It's been an exciting ride. And it's not over yet. There is a good case to be made for the proposition that the next 50 years in science will see a greater accumulation of scientific knowledge than any half-century so far, and that there are still new paradigms to be constructed.

So what is the message that comes out of the previous pages? It is that scientific advance is the combined product of:

- The time available to think and talk about science
- Opportunities for sharing ideas with other scientists
- Freedom from political, religious and cultural constraints on scientific enquiry,
  and
- The availability of appropriate technology.

On all four counts, we are better placed to seek out answers than ever before in history. Where past ages had a handful of leisured gentlemen amateurs or academics, we have hundreds of thousands of full-time paid scientists, male and female. As a result of recent developments in telecommunications – and above all, of the Internet – the opportunities for networking, and the speed of diffusion of new knowledge, far exceed anything known even a quarter of a century ago. The technology available to us, especially in computing power, is immeasurably more powerful than that available to our predecessors. And, despite a few dark corners, the freedom to pursue enquiry, and the cultural imperative to do so, are built-in characteristics of our modern world. Of course, we might just happen to be living at a time when most of what there is to be discovered has been discovered. But the history of science is littered with stories of eminent scientists who felt sure that they too were living at such a time. And how wrong they proved to be!

It has been a rather private party so far. For nearly 500 years, from Copernicus to the Human Genome Project, Europe and North America, where the money was, where the leisure was, and where the military technology was, had a virtual monopoly of science. But as we observed when we discussed the Scientific Revolution of the seventeenth century, that was not because the nations involved were cleverer than others; they were just luckier. Now, at long last, China, India, and a score of other countries have a chance to show what they can do; and the consequence can only be a further quickening of the pace of scientific advance. Hang on to your hat!

APPENDIX 1: MEASURING THINGS

VERY SMALL AND VERY LARGE NUMBERS As science had to concern itself more and more with the very large and the very small, scientists needed a language to help them handle numbers with lots of noughts, or lots of decimal places. When kilowatts became to small to cope with increasing power outputs, megawatts were used instead. When millimetres became too clumsy to handle small dimensions, micrometres were used instead. But for many modern needs, the megawatt is too small, and the micrometre is too clumsy. So scientists have settled on a standard set of prefixes, which can be attached to any unit of measurement to keep the numbers manageable:

Prefix	Symbol	Meaning
tera	T	× 1,000,000,000,000
giga	G	× 1,000,000,000
mega	M	× 1,000,000
kilo	k	× 1,000
hecto	h	× 100
deca	da	× 10
deci	d	÷ 10
centi	c	÷ 100
milli	m	÷ 1,000
micro	μ	÷ 1,000,000
nano	n	÷ 1,000,000,000
pico	p	÷ 1,000,000,000,000

These prefixes can be applied to any unit. A picosecond (ps) is a million millionth of a second, and a terawatt hour (Twh) is a thousand million kilowatt hours. When measuring length, a millionth of a metre is usually referred to as a *micron*, rather than a micrometre.

MEASURING TEMPERATURE   Temperature is not the same as heat. A kettle of boiling water and a cup of boiling water have the same temperature, but the kettle contains a lot more water, and therefore a lot more heat, which is why it can melt a lot more ice. Strange as it may sound, a bucket of ice contains a lot more heat than a cup of ice at the same temperature. Heat is lost as temperature falls. When a liquid freezes, it still contains a lot of heat, which it can lose if its temperature falls further.

But heat loss cannot continue indefinitely, because every substance contains only a finite amount of heat. At a temperature of *minus* 273 degrees Celsius, there is no heat left to lose, and cooling can go no further. Physicists call this temperature *absolute zero,* and it provides the starting point of their preferred scale – *degrees Kelvin* – for measuring temperature. There are two other scales in common use for measuring temperature: *Fahrenheit* and *Celsius* (also known as *Centigrade*).

The Fahrenheit scale was devised by the German-Dutch physicist Daniel Fahrenheit. The son of a wealthy merchant, he was born in Danzig (now Gdansk) in 1686. After his parents died, he moved to Amsterdam, and set up as a manufacturer of meteorological instruments. In 1714 he had the inspired idea of using mercury in thermometers instead of alcohol. This enabled his instruments to measure temperatures well below the freezing point of water, and well above its boiling point. So as to avoid the frequent use of negative numbers on cold days, he added salt to water to obtain a lower freezing point. He defined this as 0°. He defined the freezing point of unsalted water as 32°, and the boiling point of water as 212°. In 1724, his innovation earned him membership of the Royal Society, and his system was adopted in Britain and the Netherlands immediately afterwards.

The Celsius scale was the brainchild of a contemporary of Fahrenheit, the Swedish astronomer Anders Celsius. He was born in Uppsala, in 1701, into a famous scientific family, and became professor of astronomy in Uppsala. In 1742 he published his scheme for a new scale of temperature measurement – the 'centigrade' scale – in which the boiling point of water was defined as 0°, and its freezing point as 100°. In the following year, he reversed these figures to make freezing point 0° and boiling point 100°. This was the beginning of the centigrade system that became the standard in most non-English-speaking countries, and the preferred tool of scientists everywhere. In 1948, it was agreed that it should be renamed the Celsius scale.

Physicists nowadays commonly work in terms of *kelvins*. This commemorates the name of the British physicist William Thomson, Lord Kelvin, who introduced the concept of absolute zero. On the Kelvin scale, absolute zero – *minus 273° Celsius* – is defined as 0°, and all measurements have a positive value.

The rules for conversion from one scale to another are as follows:

Fahrenheit to Celsius
Deduct 32, divide by 9, and multiply by 5.

Example: 212°F − 32 = 180; 180 ÷ 9 = 20; 20 × 5 = 100°C.

Celsius to Fahrenheit
Divide by 5, multiply by 9, add 32.
Example: 100°C ÷ 5 = 20; 20 × 9 = 180; 180 + 32 = 212°F

Celsius to Kelvin
Add 273.
Example: 0°C + 273 = 273K

SOME INTERNATIONAL STANDARDS The SI (*Système Internationale*) definition of a **metre** is the distance travelled by light through a vacuum in $\frac{1}{299,792,458}$th of a second.

The SI definition of a **second** is 9,192,631,770 vibrations of a cesium atom, as measured by an atomic clock.

APPENDIX 2: TIMELINES

TIMELINE OF ASTRONOMY

BC

c. 3000   The Babylonians predict eclipses

c. 2400   Chinese astronomers adopt the Equatorial system of making observations, 4,000 years before its use by Tycho Brahe

c. 1800   Star catalogues and planetary records are compiled in Babyloni

c. 1600   The constellations of the Zodiac are identified in Mesopotamia

c. 530   Pythagoras asserts that the Earth is a sphere

352   Chinese astronomers record the first known sighting of a supernova

c. 270   Aristarchus of Samos calculates the distance to the Sun, and suggests that the Earth and the planets travel in orbits around

c. 260   Eratosthenes calculates the circumference of the Earth

c. 130   Hipparchus calculates the size and distance of the Moon

AD

c. 140   Ptolemy writes his *Megale Syntaxis*. His model of the solar system will remain unchallenged for 1,400 years

827   Ptolemy's book is translated into Arabic as the *Almagest*

c. 880   The Arab astronomer Al-Battani calculates the length of the yea

c. 940   A Chinese star map uses the Mercator projection

c. 1000   Ibn Yunnus of Cairo publishes the Hakimitic star tables

1175   First Latin translation of Ptolemy's *Almagest*

1543   Publication of Copernicus' *On the Revolution of the Heavenly Bodies*

1572   Tycho Brahe observes his supernova

1609   Kepler formulates his laws of planetary motion

1610   Galileo's telescope reveals that the Milky Way is made up of millions of individual stars

1672   Cassini calculates the distances of the Earth and the other planets from the Sun with an error of less than 10%

1687	Publication of Newton's *Principia*, setting out his theory of universal gravitation
1796	Laplace suggests that the solar system condensed from a cloud of gas
1814	von Fraunhofer analyses the Sun's light spectroscopically
1838	Bessel obtains the first measurement of the distance of a star
1912	Leavitt's study of variable stars provides the key to the distance of other galaxies
1912	Russell publicises his theory of stellar evolution, later embodied in the Hertsprung–Russell diagram
1916	Einstein publishes his general theory of relativity
1919	Eddington's observations of a solar eclipse provides support for Einstein's general theory
1924	Hubble resolves the arms of the Andromeda Nebula into stars, showing it to be a separate star system from the Milky way
1928	Payne discovers that stellar atmospheres are mainly composed of hydrogen
1929	Hubble interprets the red shift in the spectra of extra-galactic nebulae as evidence that the Universe is expanding
1938	Bethe and von Weizsacker propose that the Sun's heat derives from the conversion of hydrogen into helium
1948	Gamow and Alpher advance their 'big bang' hypothesis as an explanation of the origin of the universe
1965	Dicke identifies cosmic background radiation, providing support for the 'big bang' hypothesis
1987	Observations of a supernova in the Larger Magellanic Cloud confirm the accepted model of the internal processes of stars
1995	Mayor and Queloz discover a planet orbiting a star 50 light-years from the Earth

TIMELINE OF BIOLOGY

BC

c. 350   Aristotle arranges 500 species of animals in 8 classes

c. 300   Theophrastus' writings lay the foundations of botany

AD

c. 150   Galen summarizes current medical knowledge

c. 1000   The *Canon of Medicine* by Avicenna (Ibn Sina)

c. 1180   Galen is translated into Latin

1551   Gesner's *Historia animalum*: the beginning of modern zoology

1604   Fabricius publishes a comparative study of animal foetuses

1614   Sanctorius' *De statica medicina*, the first study of metabolism

1628   Harvey's *On the Motions of the Heart and the Blood*

1665   Hooke's *Micrographia*

1670   Ray pioneers species as the basis of classification

1677   Leeuwenhoek describes human sperm

1735   Linnaeus' *System of Nature*

1800   Cuvier's *Lessons in Comparative Anatomy*

1809   Lamarck's *Philosophy of Zoology*

1826   Baer describes female egg cells in mammals

1865   Darwin's *On the Origin of Species*

1865   Mendel publishes an article setting out his theories of genetics

1868   Pasteur identifies bacilli causing silkworm disease

1869   Miescher discovers DNA (which he calls 'nuclein')

1871   Darwin's *Descent of Man*

1878   Kuhnne describes enzymes

1898   Beijerincke identifies the tobacco mosaic virus

1900   Rediscovery of Mendel's work on heredity

1901   De Vries introduces the concept of mutations

1902	Sutton's *Chromosome Theory of Heredity*
1926	Sumner shows that enzymes are proteins
1932	Morgan's *The Scientific Basis of Heredity*
1934	Bernal takes first X-ray diffraction photograph of a protein crystal
1937	Dobzhanski's *Genetics and the Origin of Species*
1938	Capture of a coelacanth, thought extinct for 60 million years
1944	Avery demonstrates that DNA is the vehicle of heredity
1953	Tinbergen's *The Herring Gull's World*
1953	Crick and Watson establish the structure of the DNA molecule
1975	Wilson introduces the concept of 'sociobiology'
1977	Woese proposes *Archaea* as a third primary form of life, in addition to *prokaryotes* and *eukaryotes*
1990	Location of the cystic fibrosis gene
2001	Completion of the Human Genome Project

TIMELINE OF CHEMISTRY

BC

c. 3000   Bronze is made from copper and other metals in Egypt and the Middle East

AD

1661   Boyle distinguishes mixtures from compounds in *The Sceptical Chymist*

1669   Brand discovers phosphorous

1756   Black discovers carbon dioxide

1772   Daniel Rutherford discovers nitrogen

1773/74   Scheele and Priestley independently discover oxygen

1776   Cavendish isolates nitrogen

1784   Cavendish establishes that water is a compound of hydrogen and oxygen

1789   Lavoisier's *Elementary Treatise on Chemistry* lists 33 elements

1791   Richter establishes that acids and bases always neutralize in the same proportion

1799   Proust's Law: elements combine in fixed proportions by mass

1807   Davy uses electrolysis to isolate potassium and sodium

1808   Dalton outlines his atomic theory in his *New System of Chemical Philosophy*

1811   Avogadro coins the word molecule, and formulates Avogadro's law

         Berzelius introduces the modern system of chemical notation

1828   Woehler prepares urea from inorganic materials

c. 1835   Faraday formulates the laws of electrolysis

1845   Hofmann creates synthetic aniline

1846   Schonbein accidentally discovers gun-cotton

1852   Frankland introduces the concept of valency

1856   Perkin synthesizes the colour mauve

1858	Kekulé explains large organic molecules in terms of the four bonds of carbon
1859	Kirchhoff and Bunsen use a spectroscope to identify elements
1869	Mendeleyev publishes his periodical table
1875/76	Discovery of gallium and scandium, as predicted by Mendeleyev
1877/78	Swan and Bernigaud successively patent rayon in England and France
1887	Arrhenius provides theoretical background for Faraday's law of electrolysis
1898	Marie and Pierre Curie discover radium and polonium
1908	Haber develops a process for extracting nitrogen from the air to make ammonia
1913	Moseley formulates his law of atomic numbers
1916	Lewis explains chemical bonding and valence by his theory of shared electrons
1919	Langmuir's concentric shell model provides further elucidation of the phenomenon of valence
1923	Bronstead defines acids and bases in terms of hydrogen ions
1932	Urey discovers deuterium
1933	Segre synthesizes technetium, the first artificial element
1939	Pauling's *The Nature of the Chemical Bond*
1940	Kamen discovers carbon 14
1940	Seaborg and McMillan create the first transuranic element, neptunium
1940–85	*Fourteen* new synthetic elements are added to the Periodic Table
1985	Kroto and Smalley discover giant molecules of carbon

TIMELINE OF EARTH SCIENCE

BC

c. 570   Xenophanes interprets fossil seashells as evidence of past subsidence

AD

132   Zhang Heng constructs his earthquake detector

1517   Fracastoro suggests fossils are the remains of living creatures

1743   Packe's *New Philosophical Chart of East Kent* – the first geological map

1744   Lomonosov publishes a catalogue of 3,000 minerals

1763   Guettard and Lavoisier's mineralogical atlas of France

1770   Benjamin Franklin maps the Gulf Stream

1774   Werner proposes a formal system of classifying minerals

1779   Buffon suggests that the Earth is 75,000 years old

1785   Hutton's *Theory of the Earth*

1795   Cuvier identifies Mosasaur remains as belonging to a prehistoric giant reptile

1798   Cavendish determines the mass and density of the Earth

1799   Humboldt identifies the Jurassic epoch

1809   McLure's geological map of the eastern United States

1811   Cuvier's theory of catastrophic extinctions

1815   Smith's *Geological Map of England* defines rock strata by their fossil content

1822   Identification of the Cretaceous and Carboniferous epochs

1831   Lyell's *Principles of Geology*

1835   Identification of the Cambrian and Silurian epochs

1837   Agassiz promotes the concept of an 'Ice Age'

1842   Owen coins the word 'dinosaur'

1842   Darwin's *Structure and Distribution of Coral Reefs*

1855	Maury's *Physical Geography of the Sea* establishes the science of oceanography
1866	Daubree suggests the Earth has a nickel-iron core
1906	Oldham uses earthquake waves to establish the existence of the Earth's core
1909	Mohorovicic discovers the boundary between the Earth's mantle and core that now bears his name
1912	Wegener puts forward his theory of continental drift
	Matuyama reveals evidence of periodic reversals in the Earth's magnetic field
1935	Richter devises his scale for earthquake measurement
1953	Patterson arrives at an age for the Earth of 4.6 billion years
1954	Barghoorn and Tyler discover fossils 1.5 billion years older than any previously known
1958	Bachus and Herzenberg prove the existence of a 'dynamo' within the earth, creating its magnetic field
1961/2	Dietz and Hess propose theories of sea-floor spreading
1968	Barghoorn discovers traces of amino acids, indicative of the existence of life, in 3-billion-year-old rocks
1980	Alvarez discovers the K/T extinction boundary
1987	Measurements of continental drift by NASA satellites confirm tectonic theory
1999	Samples from Andean lakes provide 15,000-year-old profile of El Niño effects on global climate

TIMELINE OF PHYSICS

BC

c. 430 Democritus speculates that all matter is composed of indivisible atoms

c. 250 Archimedes discovers the laws governing behaviour of levers and floating bodies

AD

c. 1000 The Arab physicist Ibn al Haytham distinguishes transmitted and reflected light, explains the workings of lenses and formulates the laws of reflection

1590/91 Galileo's *On Motion* and *On Mechanical Sciences*

1621 Snell develops laws governing the refraction of light

1632 Galileo's *Dialogue Concerning the Two World Systems* is banned by the Church

1638 Galileo enunciates the principle that distance fallen increases with the square of the time elapsed

1676 Roemer uses the moons of Jupiter to measure the speed of light

1687 Newton's *Principia* makes public his theory of universal gravitation.

1690 Huygens' *Treatise on Light* describes his wave theory of light

1704 Newton's *Opticks* presents his ideas on the nature and behaviour of light

1762 Black discovers latent heat and specific heat

1798 Cavendish establishes the value of the gravitational constant

1800 Volta generates the first electric current

1803 Dalton proposes that matter consists of atoms

1811 Avogadro enunciates his law concerning the number of particle in a given volume of any gas

1820 Oersted discovers electromagnetism

1827 Ampère formulates his laws of electromagnetism

1831 Faraday and Henry independently establish that a moving magnet generates an electric current

1847	Joule and Mayer independently formulate the law of conservation of energy
1850	Clausius enunciates the second law of thermodynamics
1851	Kelvin introduces the concept of absolute zero
1873	Maxwell's *Electricity and Magnetism* creates the science of electromagnetism
1887	Michelson and Morley demonstrate that the speed of light is independent of the Earth's motion
1888	Hertz detects radio waves and measures their wavelength
1895	Röntgen discovers X-rays
1897	Thomson (Joseph J.) discovers the electron
1898	The Curies discover and name radioactivity
1900	Planck introduces the concept of quanta: the beginning of quantum theory
1905	Einstein publishes his special theory of relativity
1911	Rutherford presents his planetary model of the atom
1912	Bohr publishes his theory of the orbital behaviour of electrons
1916	Einstein completes his general theory of relativity
1924	De Broglie proposes a dual particle–wave character for atomic particles
1927	Heisenberg promulgates his uncertainty principle
1932	Chadwick discovers the neutron
1939	Meitner publishes her paper on the splitting of the uranium atom
1942	Fermi and his team achieve controlled atomic fission
1947	Feynman, Swinger, and Tomonaga independently develop quantum electrodynamics (QED), linking the behaviour of light and the behaviour of matter
1955	Atoms are directly observed for the first time
1964	Gell-Mann proposes that the heavier subatomic particles are composed of quarks
1986	Muller and Bednorz discover high-temperature superconductivity
	Individual quantum jumps in individual atoms are observed for the first time

BC

c. 1500   Thutmose III erects 'Cleopatra's Needle', used to measure tim
and the seasons

c. 1000   Alphabetical writing appears in the eastern Mediterranean

c. 132   Chang Heng's earthquake detector

AD

c. 100
(or earlier)   Paper is used for writing in China

c. 710   The earliest surviving fragment of Chinese printed text

846   The earliest surviving Chinese printed book

c. 1250   Eye-glasses of transparent quartz in use in Europe and China

c. 1310   Mechanical clocks make their first appearance in Europe

1455   Gutenberg's Bible: the first book using movable type

1569   Tycho Brahe builds his 6-metre/120-foot quadrant for stellar
observations

c. 1600   Janssen constructs the first compound microscope

1609   Galileo makes his first telescope

1657   Huygens invents the pendulum clock

1764   Harrison's H4 chronometer passes its trials

1817   von Fraunhofer develops the prism spectrometer

1827   J.J. Lister improves the compound microscope

1845   The 1.8-metre/72-inch Rosse telescope

1856   Palmieri invents the seismograph

1860   Kirchhoff establishes the foundations of stellar spectroscopy

1906   Tsvett develops paper chromatography

1907   Boltwood develops uranium-lead radioactive dating

1912   von Laue invents X-ray crystallography

1917   Completion of the 2.5-metre/100-inch Hooker (Mount Wilso
telescope

1932	The first electron microscopes
1932	Lawrence constructs a cyclotron to accelerate subatomic particles
1936	Turing lays the foundations of computing theory
1937	Reber constructs a radio telescope
1942	The world's first atomic pile is constructed at the University of Chicago
1946	The University of Pennsylvania's ENIAC computer
1947	Completion of the 5-metre/200-inch Hale (Mount Palomar) telescope
1955	The field ion microscope makes it possible to image single atoms
1956	FORTRAN: the world's first programming language
1958	The first computers using transistors instead of valves
1960	Maimen develops the first laser
1969	The Arpanet, forerunner of the Internet
1971	The first microprocessor is introduced in the United States
1988	Development of the polymerase chain reaction (PCR) technique
1990	Completion of the Hubble Space Telescope Tim Berners-Lee conceives the World Wide Web
1991	Invention of the scanning tunnelling microscope (STM)
1992	Launch of the COBE satellite

FURTHER READING

A Short History of Nearly Everything by **Bill Bryson**
(Broadway, 2003)
Bill Bryson's book is not a history of science. It is a summary of what one man learned about the present state of scientific knowledge by talking to experts. It is popular science writing at its best. Anyone who likes reading about science should have a copy.

Science: A History, 1543–2001 by **John Gribbin** (Penguin, 2002)
This *is* a history of science, and it is probably the best one-volume introduction currently available.

Landmarks in Western Science by **Peter Whitfield** (The British Library, 1999)
This beautifully illustrated book is recommended to anyone who wishes to know more about the early history of science.

On Giants' Shoulders, **edited by Melvyn Bragg** (Hodder and Stoughton, 1998)
The text of a series of radio discussions about famous scientists of the past, conducted by present-day experts on the people under discussion. The scientists discussed include Archimedes, Galileo, Newton, Lavoisier, Faraday, Darwin, Curie, Einstein, and Crick and Watson. Highly recommended.

A Brief History of Science, **by Thomas Crump** (Robinson 2002)
A history of astronomy, chemistry and physics, from the earliest times to the present day, as seen through the development of scientific instruments.

Discover Science Almanac, **edited by Bryan Bunch and Jenny Tesar** (Hyperion,2003)
A treasure-trove of facts and figures, relating to every branch of science.

INDEX